Thin Films - Growth, Characterization and Electrochemical Applications

Edited by Fatma Sarf,
Emin Yakar and Irmak Karaduman Er

Published in London, United Kingdom

Thin Films - Growth, Characterization and Electrochemical Applications
http://dx.doi.org/10.5772/intechopen.107556
Edited by Fatma Sarf, Emin Yakar and Irmak Karaduman Er

Contributors
Arturo Fernández Madrigal, Arelis Ledesma Juárez, Alejandro Altamirano Gutiérrez, Snehal Mandal, Dongmei Dong, Suman Sardar, Emmanuel Ifeanyi Ugwu, Ehsan Rahmani, Daniel Schwarz, Michael Oehme, Erich Kasper, Junjie Yang, Huiwen Deng, Jae-Seong Park, Siming Chen, Mingchu Tang, Huiyun Liu

First published in London, United Kingdom, 2024 by IntechOpen
IntechOpen is the global imprint of INTECHOPEN LIMITED, registered in England and Wales, registration number: 11086078, 5 Princes Gate Court, London, SW7 2QJ, United Kingdom

British Library Cataloguing-in-Publication Data
A catalogue record for this book is available from the British Library

Additional hard and PDF copies can be obtained from orders@intechopen.com

Thin Films - Growth, Characterization and Electrochemical Applications
Edited by Fatma Sarf, Emin Yakar and Irmak Karaduman Er
p. cm.
Print ISBN 978-1-83769-135-7
Online ISBN 978-1-83769-136-4
eBook (PDF) ISBN 978-1-83769-137-1

For EU product safety concerns:
IN TECH d.o.o., Prolaz Marije Krucifikse Kozulić 3, 51000 Rijeka, Croatia,
info@intechopen.com or visit our website at intechopen.com.

Meet the editors

Assoc. Prof. Fatma Sarf received her master's degree from Konya Selçuk University, Turkey, in 2007 and her Ph.D. from Canakkale Onsekiz Mart University, Turkey, in 2016. Dr. Sarf has investigated the optical properties of metal oxides, surface wettability, electrode materials for Li-ion batteries, and metal oxide gas sensors. She has produced thin films and nanoparticles using different systems. Recently, she has focused on waste materials as anode materials for Li-ion batteries and supercapacitors.

Assist. Prof. Dr. Emin Yakar received his master's degree and Ph.D. from the Gebze Institute of Technology, Turkey. He has investigated experimental and theoretical studies of semiconductor materials. He has focused on ceramics, surface wettability, and the characterization of metal oxide thin films. Dr. Yakar also studies metal oxide gas sensors, ceramics, and phase transitions in semiconductors.

Assoc. Prof. Irmak Karaduman Er received her Ph.D. from Gazi University, Turkey, in 2015. She has investigated the optical properties and gas sensor characteristics of semiconductors. She produced thin films using chemical bath deposition and other chemical techniques. Dr. Karaduman is currently also studying the photoluminescence of different kinds of phosphorous materials.

Contents

Preface

Thin films with and without plasticizers defined by Tian et al. (2020), usually known as thin and/or elastic polymers, are indispensable in technology. Homo or hetero structures based on thin films have attracted great attention with the use of different chemical or physical production techniques. In this way, advanced functional device design and processes up to the atomic level can be achieved at all length scales.

The idea of writing this book developed as a result of interest in the fields of electrochemical production and applications. The book is divided into four sections on the production and characterization of thin films, electrochemical applications, electrodeposition, and magnetic applications.

This book was created with the help of the co-editors Assistant Professor Emin Yakar and Associate Professor Irmak Karaduman Er. We express our thanks to all authors, referees, and advisors for their help and support in making the publication of this book possible.

Fatma Sarf
Can Vocational School,
Canakkale Onsekiz Mart University,
Canakkale, Turkiye

Emin Yakar
Çanakkale Onsekiz Mart University,
Canakkale, Turkiye

Irmak Karaduman Er
University of Çankırı Karatekin,
Çankırı, Turkiye

Section 1

Thin Films

Chapter 1

Preparation and Characterization of Thin Films by Sol-Gel Method

Ehsan Rahmani

Abstract

The sol-gel method has been widely used to prepare several materials, such as glass fibers, catalysts, electrochemical devices, or thin films. Sol-gel is considered an economical and straightforward method compared to physical vapor deposition (PVD) or chemical vapor deposition (CVD), which are more complex and need more facilities. At the same time, almost the same quality has been evaluated for sol-gel thin films. Furthermore, chemical tailoring of raw materials to produce new functional compositions is more feasible than conventional methods such as PVD. Thin films utilizing sol-gel were prepared by dip coating, spin coating, electrochemical coating, and spray coating methods, where these methods can be used for various substrate types. Prepared thin films may be utilized in several areas of application, such as semiconductors, catalysts, or photocatalysts.

Keywords: thin film, sol-gel, electrodeposition, dip coating, spin coating, spray coating, characterization

1. Introduction

Sol-gel processing began by Ebelman and Graham, who studied silica gels as early as the mid-1800s. Hydrolysis of tetraethyl orthosilicate (TEOS), $Si(OC_2H_5)_4$, under acidic conditions, yielded SiO_2; these early investigators observed that it was in the form of a "glass-like material" [1].

Sol-gel method can be implemented in several fields of scientific and engineering fields, such as the ceramic industry, nuclear industry, and the electronic industry or the development of new materials for catalysis, membranes, chemical sensors, photochromic applications, fibers, optical gain media, and solid-state electrochemical devices [2–13]. Thin film and powder catalysts have been widely produced using the sol-gel method. Several variants and manipulations have been implemented into the process to synthesize pure thin films or powders in large homogeneous concentrations and under stoichiometry control [14–16]. Process simplicity, optical complex shapes, uniform oxide complexes, special shapes of fibers and aerogels, synthesis of amorphous minerals, synthesis of porous material with high content of organic and polymeric compounds, and synthesis of highly optical transparent or low initial investment are the most advantages of the sol-gel method [17–19].

The term sol is attributed to a stable suspension containing colloidal nanoparticles (with diameters of 1–100 nm). The sol may contain amorphous or crystalline particles with dense, porous, or polymeric substructures [20, 21].

A rigid network with pores of sub-micrometer dimensions and polymeric chains whose average length is greater than a micrometer has been considered a gel [22]. A diversity of combinations of substances embraced as the "gel" can be classified into four categories: (1) structures of well-ordered lamellar; (2) networks of covalent polymer that is completely disordered; (3) formation of predominantly disordered through physical aggregation of polymer networks, (4) disordered particular structures [23–25].

The gel is attributed to a porous three-dimensional continuous solid network surrounding and supporting a continuous liquid phase. Synthesis of oxide materials and gelation is because of covalent bond formation between the sol particles [26, 27]. Gel formation can be reversed due to the presence of van der Waals interactions or hydrogen bonds. The gel structure network is dependent on the size and shape of the sol particles to a large extent [28].

2. Preparation of thin film by sol-gel method

Sol-gel monoliths have been prepared by three approaches: method 1, colloidal powders solution; method 2, alkoxide or nitrate precursors hydrolysis and polycondensation followed by hypercritical drying of gels; method 3, aging and drying of alkoxide precursors under ambient atmospheres after hydrolysis and polycondensation [29, 30].

The hydrolysis and condensation reactions (sol-gel process) would be influenced by several parameters, such as the metal alkoxide activity, solution pH, ratio of the water/alkoxide, temperature, solvent nature, and additive used. The addition of catalysts controls the rate, hydrolysis extent, and condensation reactions. By varying processing parameters, materials with different microstructures and surface chemistry can be obtained [31]. The fabrication of ceramic materials in various forms would be available by further processing the "sol." By casting the sol into the mold, a wet gel will form. Dense ceramic or glass particles will form, followed by drying and heat treatment of the gel [32]. Meanwhile, highly porous and low-density aerogel material is synthesized utilizing supercritical conditions to remove the liquid in a wet gel. Many natural systems like opals, agates, and particles are evidence of silicate hydrolysis and condensation to form polysilicate gel. Ebelman prepared the first metal alkoxide from $SiCl_4$ and alcohol and found the first "precursor" for glassy materials can be $Si\text{-}(OC_2H_5)_4$ and gelled compound on exposure to the atmosphere [33].

$$\text{Step 1: (Hydrolysis): } Si(OR)_4 + H_2O \leftrightarrow SiOH(OR)_3 + R{-}OH \tag{1}$$

$$\text{Step 2: (Water condensation): } SiOH(OR)_3 + SiOH(OR)_3 \leftrightarrow [Si]_2O(OR)_6 + H_2O$$

$$\text{(Alcohol condensation): } SiOH(OR)_3 + Si(OR)_3 \leftrightarrow [Si]_2O(OR)_6 + R{-}OH) \tag{2}$$

2.1 Electrochemical aided deposition (EAD)

Dip coating, spin coating, and spraying methods are usually used to prepare sol-gel films. Only flat surfaces can be coated by dip coating and spin coating, while the rheology of the precursor and fine-tuning must be considered for the spray coating method. Sol-gel thin film preparation implementing the electrochemical deposition technique has been considered in the last two decades. Woo et al. applied this technique for preparing silane films as binders [34]. Electrodeposition of silane-based films and proposed its mechanism later studied by Shacham et al. [35]. Preparation of coatings on conductive materials based on the sol-gel principle utilizing EAD is a recently developed method, and thicker and rougher sol-gel films can be prepared by this method. Application of negative potential and in-situ catalysis of sol-gel chemistry (hydrolysis and condensation reactions) are the basic principles of EAD. Reduction of oxygen or some specific ions supporting the electrolyte or hydrolysis, which leads to an increase in pH value near the electrode, would accelerate the immediate condensation of the solute on the electrode surface. Equation (3) expressed the chemical reaction in this process [36],

$$O_2 + 2H_2O + 4e \rightarrow 4OH^{-(3)} \tag{3}$$

During the electrodeposition process, the pH value of the bulk solution does not change, and no new materials are introduced; the silane component does not lose or gain electrons at the electrode surface. Electrodeposited films have several unique advantages compared to conventional self-assembled films [37]:

1. The base catalyst provides an additional driving force for the film, resulting in thicker and rougher film formation near the cathode surface.

2. Films with greater porosity and better intra-cohesion may be produced due to gelation, where solvent evaporation occurs during electrodeposition.

3. The derived OH-ions can catalyze the chemical bonding process between the sol-gel film and the substrate.

Coatings on complex non-planar geometries and controlling the thickness and composition of the nanocomposite are the significant advantages of depositing utilizing EAD (**Figure 1**).

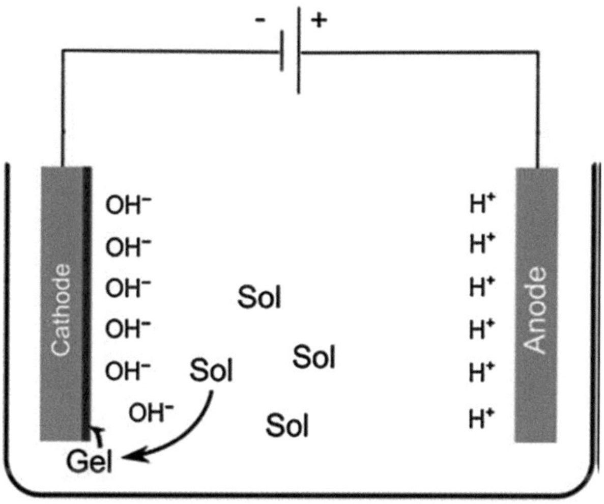

Figure 1.
Electrochemical deposition of a sol-gel film [36].

2.2 Dip coating method

Sol-gel dip coating consists of depositing a solid film by withdrawing a substrate from a sol: gravitational draining and solvent evaporation, followed by further condensation reactions. Sol-gel dip coating, compared to methods such as chemical vapor deposition (CVD), evaporation, or sputtering, does not require complex equipment and is potentially less expensive than conventional thin film-forming processes [38]. The ability to tailor the microstructure of the deposited film is the most crucial advantage of sol-gel over conventional coating methods.

The substrate from the liquid bath is drawn vertically at a speed of U_o. By entraining the moving substrate, the liquid in a fluid mechanical boundary layer may separate above the liquid bath surface, and the outer layer returns to the bath. The fluid film terminates at a well-defined drying line by solvent evaporating and draining. The process is steady with respect to the liquid bath surface when the receding drying line velocity equals U_o.

The concomitant draining, evaporation, and hydrolysis consolidation step represents the sol-gel transition. An integrated gel film will be left due to a withdrawn drying line moving downwards with colorful parallel interference lines. By implementing volatile solvents, in comparison to the bulk sol-gel process, the complete transition will be done in a short time. The drying and keeping of the water content are almost constantly enhanced due to the evaporation and the resulting cooling. In addition, over the surface of the wet film, a downward laminar flow of vapors forms. Inhomogeneities deposition in the film properties may occur by any turbulence or variation in the atmosphere (**Figure 2**).

A fluid mechanical equilibrium between the entrained film and the receding coating liquid is the film formation process. The equilibrium state will be affected by forces such as viscous drag and the gravity force and other forces like the surface tension, the inertial force, or the disjoining pressure.

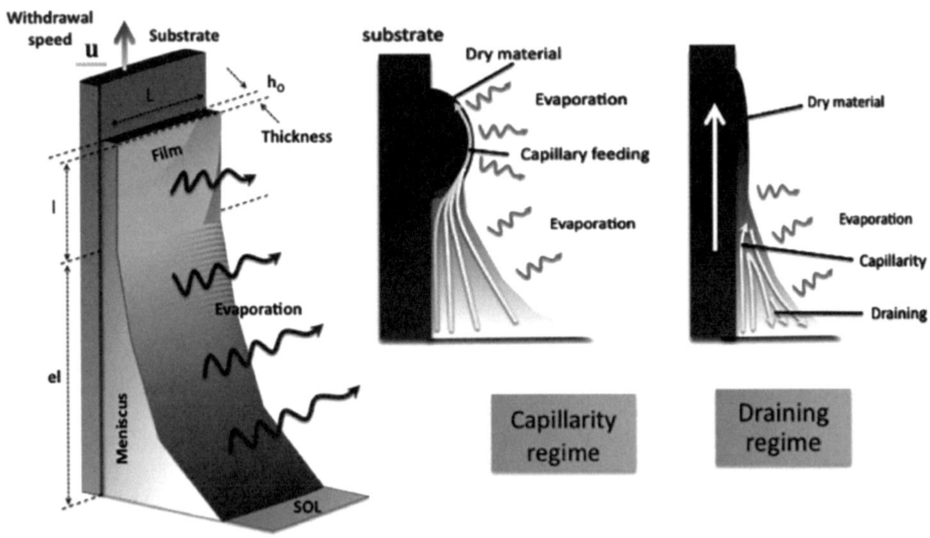

Figure 2.
Schematic of sol-gel dip coating method including both capillarity and draining regimes [39].

The Landau-Levich equation describing the final liquid film thickness *h* for pure liquids considers this fundamental theoretical approach [40],

$$h = c \cdot \frac{(\eta U)^{2/3}}{\gamma^{\frac{1}{6}}(\rho g)^{\frac{1}{2}}} \tag{4}$$

where c, η, U, y, and ρ are constant, liquid viscosity, withdrawal speed, the surface tension of the liquid against air, and liquid density, respectively.

Modified dip coating techniques can be used for more complex geometries, although this method is particularly suited for coating flat or rod-shaped rigid substrates [41]. A semi-continuous dip coating process for endless flexible substrates like webs or filaments, while dip coating is a typical batch process. Double-sided coating of flat substrates, especially in the production of optical filters, is one of the advantages of the dip coating method. For oxide coatings prepared from the metal salt solution, a single-layer film thickness can be deposited ranging from only a few nanometers to approximately 200 nm. With inorganic-organic hybrid materials, due to the lower shrinkage and the higher flexibility of the network film, several microns are accessible, and colloidal systems can be implemented to produce thicker films up to 1 μm.

The high-volume industrial production of ordinary optical filters using the dip coating technique. Three-layer antireflection coatings for technical glasses (e.g., displays and lighting) form the largest market segment. Dip coating methods have been implemented by several industrial brands such as the TiOr-based solar-control glass Calorex® (Irox®) and the antireflection coatings Amiran®, ConturanQ, and Mirogard®Several and well-known products emerged from these activities [41].

2.3 Spin coating method

The spin coating can rapidly deposit thin layers onto relatively flat substrates. The target surface is dispensed by coating solution; the spinning action causes the solution to spread out and leave behind the coated surface of the substrate by the chosen material very uniformly that is held by some rotatable fixture (often using a vacuum to clamp the substrate in place).

The viscous drag force precisely balances the rotational accelerations within the solution. Emslie, Bonner, and Peck (EBP) [I] first described this flow condition. Simultaneously, solvent evaporation out of the top surface of the solution is also considered by Meyerhofer [42]. Spin coating runs into two stages: viscous flow controlling and evaporation controlling. Prediction of the final coating thickness, h_f, in terms of several key solution parameters, is according to [42]:

$$h_f = x\left(\frac{e}{2(1-x)K}\right)^{\frac{1}{3}} \tag{5}$$

where e, K, and x are the solution's evaporation and flow constants and effective solids content, respectively. The evaporation and flow constants are defined as:

$$e = C\sqrt{\omega} \tag{6}$$

$$K = \frac{\rho\omega^2}{3\eta} \tag{7}$$

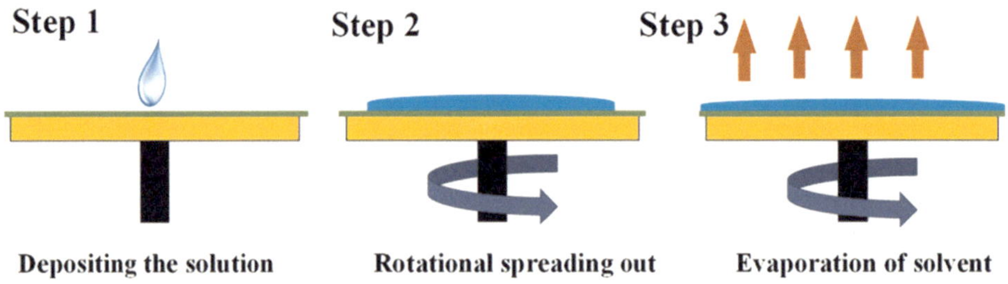

Figure 3.
Procedure for spin coating method [43].

where ω, ρ, η, and C denote the rotation rate, solution's density, viscosity, and a proportionality constant. In which C depends on airflow flow regime (laminar or turbulent) and solvent molecules diffusivity in air. Sol-gel film preparation by spin coating deposition typically results in a coating thickness below 1 μm (**Figure 3**).

2.4 Spray coating method

Spray coating is suitable for non-flat samples, and the thickness can be very well controlled and surface modification over large areas. Only a few studies have studied sol-gel spray coating.

Most of the previously reported research has considered the sol-gel dip coating method low-cost and accessible. However, it is not easy to control the homogeneity and thickness of the coating over the entire sample length in the case of dip coating. Furthermore, spray coating is suitable for non-flat samples, and the thickness can be very well controlled, a better alternative as it allows surface modification over large areas. A three-step process is a typical example of an organically modified silica hydrophobic coating prepared by spraying: (1) preparation of hydrophobic alcosol, (2) alcosol spraying on 100°C glass substrates, and (3) trimethylchlorosilane surface modification created by Mahadik et al. [44]. Due to low cost and lack of specialized equipment, traditional spray coating is still the most often reported method. At the same time, there are other procedures, including spray coating with plasma, thermal spray, and powder [45–47].

3. Characterization techniques

3.1 Thin film X-ray diffraction

Thin films have been considered the basic materials for modern electronic devices of metallic conductors, semiconductors, and insulators. These films should possess specific mechanical, electrical, magnetic, or optical properties for optimal device performance, which are affected strongly by the film's microstructural nature, such as crystalline or amorphous state, crystallographic orientation, crystallite size, strains, and stresses [48, 49]. Therefore, for the design and improvement of electronic devices, it is very important to characterize microstructural thin films.

The microstructure of thin films cannot be easily characterized by methods developed for bulk materials due to their small dimensions perpendicular to the surface. X-ray diffraction (XRD) is especially suitable for thin films since it is nondestructive, noncontact, and highly quantitative among the analytical methods.

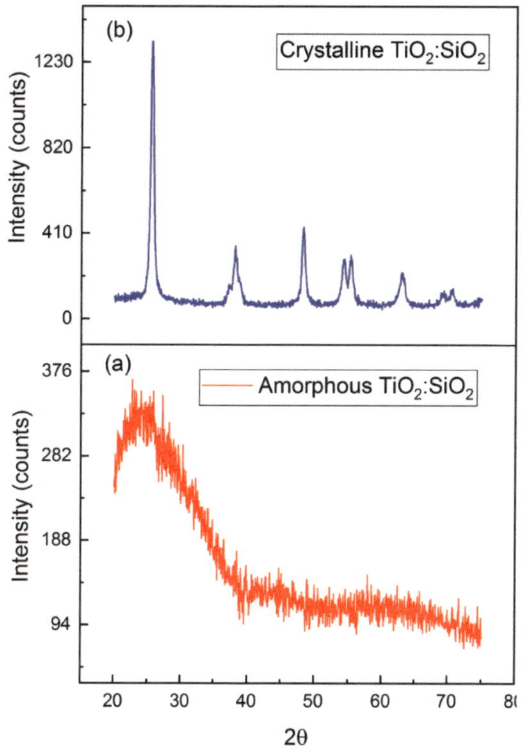

Figure 4.
XRD pattern for prepared sol-gel dip coating thin film (a) amorphous TiO_2:SiO_2 and (b) highly crystalline TiO_2: SiO_2.

By implementing the XRD test, several thin film features can be evaluated. **Figure 4** indicates XRD patterns of highly crystalline and amorphous films prepared by the sol-gel dip coating method [50], where results indicate the formation of an amorphous TiO_2:SiO_2 thin film (**Figure 4a**) and a highly crystalline TiO_2:SiO_2 (**Figure 4b**) thin film on a glass substrate. XRD results revealed that TiO_2 crystals may not be grown by an increase in SiO_2 content of more than 20 mol percent, while highly crystalline TiO_2 has formed on the glass substrate by a lower content of SiO_2 (about 15 mol%). Furthermore, by addition of SiO_2 crystalline size and crystallinity of TiO_2 declined, where Scherrer or Modified Scherrer equations have been used to evaluate the crystalline size according to XRD pattern and full width at half maximum (FWHM, β) calculation [51]:

$$D = \frac{K\lambda}{\beta \cos(\theta)} \tag{8}$$

Or modified Scherrer equation,

$$\ln \beta = \ln \frac{K\lambda}{L} + \ln \frac{1}{\cos \theta} \tag{9}$$

In addition, the crystalline phase of the synthesized thin film can be indicated by considering XRD results. **Figure 5** illustrate the XRD pattern of TiO_2 nanocrystalline thin film prepared by different heat treatment. As shown, with an increase in

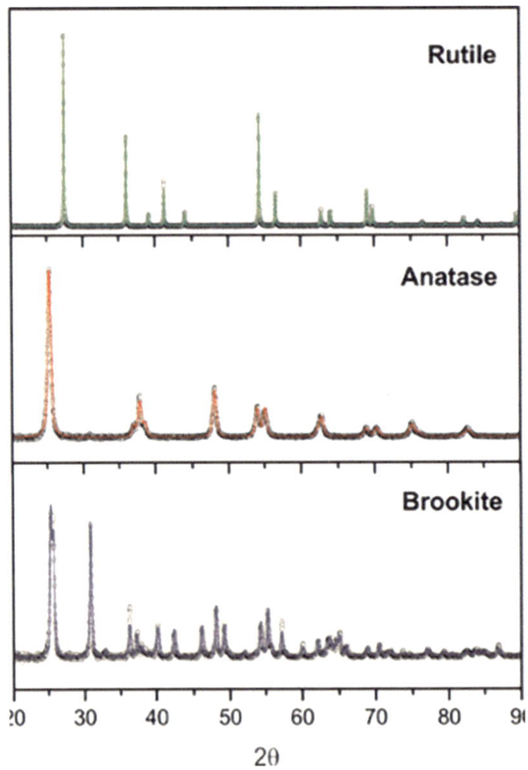

Figure 5.
Phase transmission of TiO_2 from Anatase to Brookite Phase [52].

temperature, the crystalline phase has been transformed from Anatase to Brookite phases. TiO_2 can be prepared in three phases by different heat treatment methods, where there is the Anatase phase at a heat treatment temperature below 500°C. In contrast, the Rutile phase will form around 500–750°C, and the Brookite phase formation temperature is approximately 750–900°C.

3.2 Energy dispersive x-ray spectroscopy (EDX)

Investigation of the chemical species present in a material can be evaluated by energy-dispersive X-ray spectroscopy as a powerful technique. Analysis of the energy and intensity distribution of the X-ray signal produced by the interaction of an electron beam with a specimen has been considered the basis of the EDX method. Directed an electron beam toward the sample has been analyzed in a setup that EDX is an optional tool installed on electron microscopes. The electrons can interact with the nucleus or a specific atom's electrons by certain kinetic energy. The detection of the emitted radiation due to the loss of any amount of energy of electrons between zero and the initial energy will lead to a continuous electromagnetic spectrum that constitutes the background of the collected spectrum [53].

Ionization phenomena usually involve K shell electrons when the primary electrons interact with the atom's electrons. The electrons from L or M shells can occupy the vacancy left by the K shell electron because an atom in the excited state tends to reach the minimum energy. X-ray radiation emission is designated as K_α, and K_β

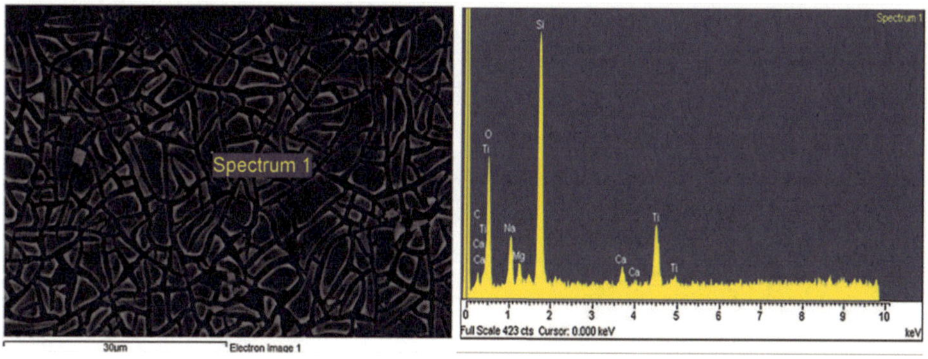

Figure 6.
EDX results for prepared TiO_2:SiO_2 thin film by dip coating sol-gel method [50].

results from a transition between L and K or M and K shells. The emission of K_α and K_β X-rays identifies the elements in the analyzed sample because of the energy difference between L, M, and K levels that are well-defined for each element. In principle, a line in the continuous electromagnetic spectrum should be given by the detection of this transition level [54].

Figure 6 shows EDX results for TiO_2:SiO_2 thin film coated on the glass substrate, where indicated the presence of Ti, Si, and O in the film. Also, the EDX test revealed the presence of Na, Mg, and Ca that can be related to the glass substrate due to the passing of the X-ray over the prepared thin film and the detection of glass substrate elements. EDX test can be used for mapping and evaluating the synthesized film's elements distribution.

3.3 X-ray photoelectron spectroscopy

X-ray photoelectron spectroscopy (XPS) is widely implemented in surface science techniques—an X-ray based on the photoelectric effect photon absorbed by a core or valence electron. The electron will be emitted by the larger incident photon energy than the binding energy [55]. A hemispherical electron energy analyzer spectrometer collected the emitted electrons by an electrostatic lens and performed energy analysis. The binding energy E_k of the electrons can be determined from the known photon energy $h\nu$ and the measured kinetic energy E_B [56],

$$E_k = h\nu - E_B - \phi \tag{10}$$

with ϕ as the spectrometer work function. In an XPS experiment, typical incident photon energies range from several 10s to well over 1000 eV, where the same order of kinetic energies will be considered.

The X-ray penetration depth into the sample is of the order of hundreds of nanometers or more, while the kinetic energy range used in conventional XPS studies, atoms in the sample photoelectrons elastic and inelastic interactions limited to the probe depth to a maximum of a few nanometers.

A variety of features is shown in the XPS survey spectrum in **Figure 7a**, where results indicate photoemission lines of Ga (Ga 2p1/2, Ga 2p3/2, Ga3s, Ga3p, and Ga 3d), Fe (Fe 2p1/2, Fe 2p3/2, and Fe 3p), and oxygen (O 1 s). The spin-orbit components of Ga 2p3/2 and 2p1/2 have been indicated by splitting 26.9 eV, along with a loss

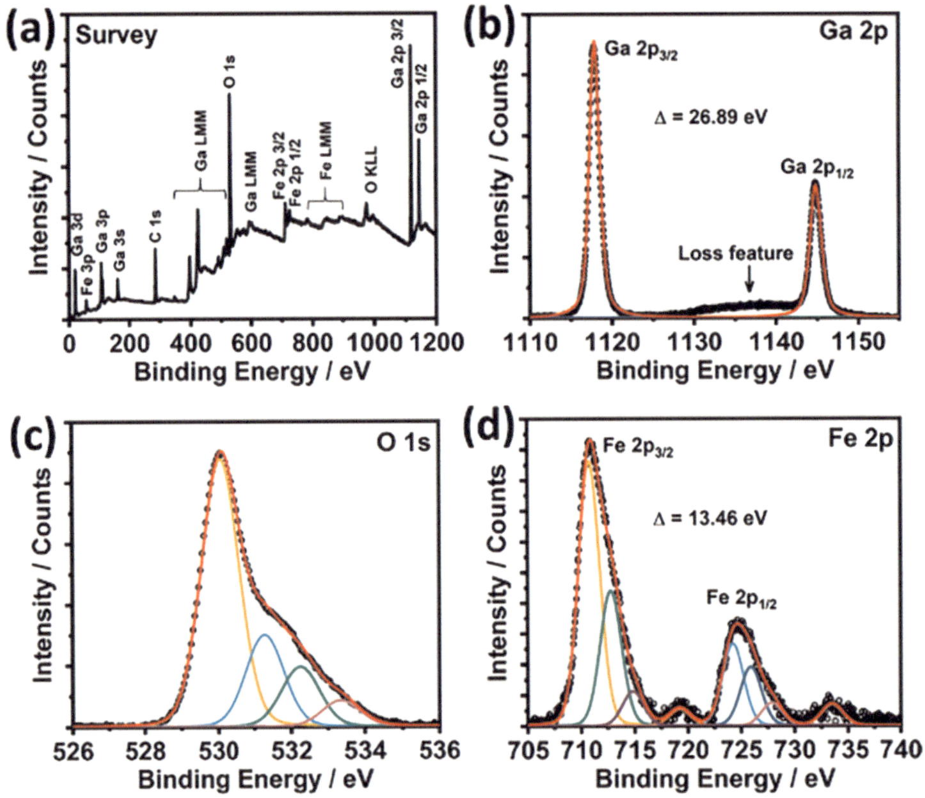

Figure 7.
GFO thin films XPS analysis, (a) different photoemissions features; (b) Ga2p core-level; (c) O1s core-level; (d) Fe2p core-level [57].

feature at around 1136 eV Ga2p (**Figure 7b**). In addition, Ga is in a +3 oxidation state confirmed by Ga 2p3/2 peak position at 1118 eV, which is characteristic of native gallium oxide [57].

Lattice-bound oxygen is assigned by the O 1 s spectrum (**Figure 7c**) centered at 530.01 eV, while surface hydroxyl and carbonylated groups are attributed to the other contributions. The component with the most intense in 2p3/2 is at 710.8 eV, with a 13.5 eV splitting from the 2p1/2 component consistent with a Fe+3 oxidation state.

3.4 Atomic force microscopic method

An amazing technique with unprecedented resolution and accuracy is atomic force microscopy (AFM), which allows us to see and measure surface structure. The arrangement of individual atoms in a sample or the structure of individual molecules can be studied by atomic force microscope imaging. The hopping of individual atoms from a surface has been measured by scanning in an ultrahigh vacuum at cryogenic temperatures. However, even seeing biological reactions occur in real-time can be carried out in physiological buffers at 37°C. AFM does not need to be carried out under these extreme conditions [58, 59]. The crystallographic structure of materials can be used to have images containing only 5 nm of microscopic images to show about 50 atoms, or 100 μm or larger images. Almost any sample, such as the surface of a

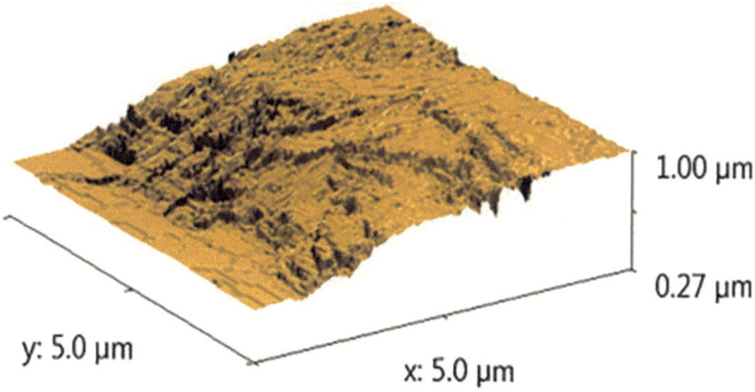

Figure 8.
AFM image for TiO_2:SiO_2 thin film prepared by sol-gel method [61].

ceramic material, highly flexible polymers, human cells, a dispersion of metallic nanoparticles, or individual molecules of DNA, can be imaged by AFM, but it is challenging [60]. Furthermore, AFM has various "spectroscopic" modes that measure other properties of the sample at the nanometer scale. A map of the height or topography of the surface building up as scanning a probe of the AFM goes along the sample surface.

Figure 8 illustrates the AFM image of a synthesized 15-layer TiO_2:SiO_2 thin film prepared by the sol-gel method. The roughness of the prepared TiO_2:SiO_2 thin film has been presented in **Figure 5**, while this roughness belongs to the crystalline formation of TiO_2 in the presence of amorphous SiO_2. In this work, TiO_2:SiO_2 gel was coated on a wall of an annular photoreactor to remove oily contamination from wastewater. However, according to AFM results, TiO_2:SiO_2 thin film roughness has been measured as 16 nm, where the film's roughness is necessary to have proper contact between the reactant and catalyst [61].

3.5 Scanning electron microscope

The scanning electron microscope (SEM) has studied surface details of several compositions, such as metals, rock minerals, polymers, corrosion deposits, filters, ceramic membranes, foils, fractured/rough surfaces, alloys, and biological samples. Conductive or non-conductive material, either in solid or powder form, can be examined in an as-received or prepared condition by SEM technique [62, 63]. A field emission gun is installed on the SEM to investigate surface features only 1 nm apart. SEM allows large areas of a sample to remain in focus at one time, yielding 3D characteristics due to its extraordinary ability to depict large depths of field.

The performance of an electron source is expressed by two essential parameters: current density and brightness. Beam current density J_b is given as [64]:

$$J_b = \frac{\text{Beam current}}{\text{Area}} = \frac{i_b}{\pi} \tag{11}$$

where i_b and d are the current and diameter of the beam, respectively, the brightness (β) of the electron source would be considered as a function of the total number of electrons (current, I) emitted from a unit area (A) of the source and the solid angle

(Ω) of emission subtended by those electrons. This relationship can be expressed as follows [65]:

$$\beta = \frac{\text{Current}}{\text{area} \times \text{solidangle}} = \frac{I}{A\Omega} = \frac{j_c}{\pi a^2} \tag{12}$$

where j_c and α are the current density (expressed in A/cm^2) and convergence angle of the beam in radians, respectively.

The SEM method has been used for analyzing thin films, where surface morphology and layer thickness can be evaluated by this method. TiO_2:SiO_2 SEM micrographs are illustrated in the **Figure 9**, where the surface morphology of 10 layers of TiO_2 and TiO_2:SiO_2 films can be observed in **Figure 9** [61]. As indicated, films have been prepared by fractured morphology utilizing the sol-gel method. Crack formation occurs through stress and different thermal coefficients of expansion of the overlayer due to contraction and substrate during the drying and annealing processes of the films. Micro-cracks will result in better diffusion and may contribute to a higher surface area while reducing the film's durability. According to SEM images, the crystalline size of TiO_2 doped by SiO_2 is 32 nm [61].

3.6 Transmitting electron microscope

Ernst Ruska and Max Knoll, in 1931, invented the transmission electron microscope. The electron microscope resolution has improved steadily from around 100 nm in the early models to 0.1 nm and is even better nowadays [66]. High-resolution transmission electron microscopy (HRTEM) refers to phase-contrast imaging resolving single atoms or atomic clusters. A highly coherent source, a well-aligned system, a good detector, and a suitable sample are essential to achieve HRTEM [67].

Samples of electron microscopy typically have gratings behavior. Most electrons travel through the sample unchanged, and none may be absorbed. Wave interference forming A TEM image.

HRTEM image is formed in the image plane when two or more selected Bragg reflected beams interact (interfere) by a suitably large objective aperture to form an image. HRTEM imaging is a type of phase-contrast imaging [68]. As a result of the

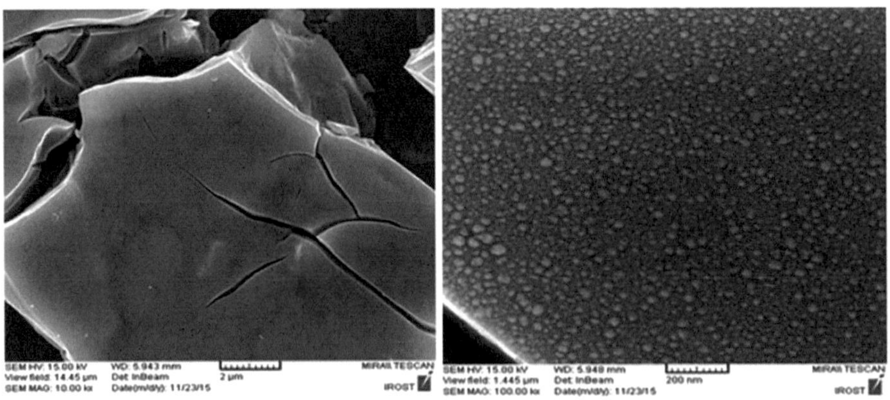

Figure 9.
SEM images for TiO_2:SiO_2 multilayer thin film [61].

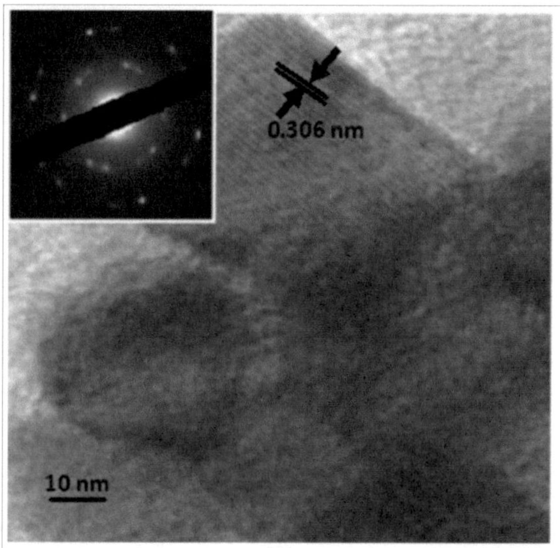

Figure 10.
TEM images for $AgInSe_2$ thin film prepared by sol-gel method [70].

contrast that occurs from the distinction in the phase of the beams as a result of their interaction with the sample, it can be used to resolve the crystalline lattice, columns of atoms, or sub-Angstrom imaging of the lattice and even single atoms can be prepared in the case of the most modern aberration-corrected transmission electron microscopes. Second-phase or amorphous layers and atomic resolution, crystalline defects, and structure across boundaries can be observed by HRTEM imaging. Furthermore, in the case of interfaces in multilayer thin films, the topography information that it is aligned correctly in the direction of the electron beam on the interface is also provided [69]. Analysis of HRTEM images is complicated and can often require simulation to directly interpret the contrast present in the image, although it is relatively easy to obtain.

Sol-gel spin coating technique has been implemented to prepare highly stoichiometric $AgInSe_2$ thin films on a p-type Si(111) substrate. Different temperatures were used to anneal these films. HRTEM images of the synthesized and annealed thin films have been shown in **Figure 10**. The HRTEM image order of the lattice spacing was indicated as 0.3 nm. Films were indexed to a pure polycrystalline chalcopyrite $AgInSe_2$ structure, as seen in the selected area electron diffraction patterns of the $AgInSe_2$ [70].

4. Conclusion

In this chapter, sol-gel method and coating techniques such as dip coating, spin coating, or electrochemical deposition have been investigated. Sol-gel coating methods have been considered economical and more feasible than methods such as physical vapor deposition

(PVD) or CVD, in which sol-gel has been widely utilized industrially. In addition, characterization methods for thin films have been discussed. X-ray-based methods (XRD, EDX, or XPS) were used to investigate the crystallinity and composition of the prepared thin film. Also, AFM, SEM, and TEM, which have been used to study the surface and structure of the films, were discussed in detail.

Author details

Ehsan Rahmani
Department of Chemical Engineering, Amirkabir University of Technology (Tehran Polytechnic), Tehran, Iran

*Address all correspondence to: ehsanrahmani@hotmail.com

References

[1] Brinker CJ. Hydrolysis and condensation of silicates: Effects on structure. Journal of Non-Crystalline Solids. 1988;**100**(1):31-50. DOI: 10.1016/0022-3093(88)90005-1

[2] Scherer GW, Brinker CJ. Acknowledgments. In: Brinker CJ, Scherer GW, editors. Sol-Gel Science. San Diego: Academic Press; 1990. pp. xiii-xiv

[3] Schubert U. Catalysts made of organic-inorganic hybrid materials. New Journal of Chemistry. 1994;**18**(10): 1049-1058

[4] Blum J, Rosenfeld A, Gelman F, Schumann H, Avnir D. Hydrogenation and dehalogenation of aryl chlorides and fluorides by the sol-gel entrapped RhCl3–Aliquat 336 ion pair catalyst. Journal of Molecular Catalysis A: Chemical. 1999;**146**(1-2, 122):117

[5] Banet P, Cantau C, Rivron C, Tran-Thi T-H. Nano-porous sponges and proven chemical reactions for the trapping and sensing of halogenated gaseous compounds; Le piegeage et la detection de composes halogenes gazeux. Utilisation d'eponges nanoporeuses et de reactions chimiques, Actualite Chimique. 2009

[6] Lin J, Brown CW. Sol-gel glass as a matrix for chemical and biochemical sensing. TrAC Trends in Analytical Chemistry. 1997;**16**(4):200-211

[7] Xomeritakis G, Tsai C, Jiang Y, Brinker C. Tubular ceramic-supported sol–gel silica-based membranes for flue gas carbon dioxide capture and sequestration. Journal of Membrane Science. 2009;**341**(1–2):30-36

[8] Zeng Z, Qiu W, Yang M, Wei X, Huang Z, Li F. Solid-phase microextraction of monocyclic aromatic amines using novel fibers coated with crown ether. Journal of Chromatography A. 2001;**934**(1–2):51-57

[9] Gvishi R, Narang U, Ruland G, Kumar DN, Prasad PN. Novel, organically doped, sol–gel-derived materials for photonics: Multiphasic nanostructured composite monoliths and optical fibers. Applied Organometallic Chemistry. 1997;**11**(2):107-127

[10] Levy D, Esquivias L. Sol–gel processing of optical and electrooptical materials. Advanced Materials. 1995; **7**(2):120-129

[11] Dunn B, Zink JI. Optical properties of sol–gel glasses doped with organic molecules. Journal of Materials Chemistry. 1991;**1**(6):903-913

[12] Levy D. Recent applications of photochromic sol-gel materials. Molecular Crystals and Liquid Crystals Science and Technology. Section A. 1997; **297**(1):31-39

[13] Dunn B, Farrington GC, Katz B. Sol-gel approaches for solid electrolytes and electrode materials. Solid State Ionics. 1994;**70**:3-10

[14] Murugan K, Subasri R, Rao T, Gandhi AS, Murty B. Synthesis, characterization and demonstration of self-cleaning TiO2 coatings on glass and glazed ceramic tiles. Progress in Organic Coatings. 2013;**76**(12):1756-1760

[15] Akpan U, Hameed B. The advancements in sol–gel method of doped-TiO_2 photocatalysts. Applied Catalysis A: General. 2010;**375**(1):1-11

[16] Malengreaux CM, Léonard GM-L, Pirard SL, Cimieri I, Lambert SD,

Bartlett JR, et al. How to modify the photocatalytic activity of TiO2 thin films through their roughness by using additives. A relation between kinetics, morphology and synthesis. Chemical Engineering Journal. 2014;**243**:537-548

[17] Shen Y, Zhao P, Shao Q. Porous silica and carbon derived materials from rice husk pyrolysis char. Microporous and Mesoporous Materials. 2014;**188**:46-76

[18] Asadpour S, Kooravand M, Asfaram A. A review on zinc oxide/poly (vinyl alcohol) nanocomposites: Synthesis, characterization and applications. Journal of Cleaner Production. 2022;**362**:132297

[19] Azadani RN, Sabbagh M, Salehi H, Cheshmi A, Raza A, Kumari B, et al. Sol-gel: Uncomplicated, routine and affordable synthesis procedure for utilization of composites in drug delivery. Journal of Composites and Compounds. 2021;**3**(6):57-70

[20] Mao N, Du M. 10 - Sol–gel-based treatments of textiles for water repellence. In: Williams J, editor. Waterproof and Water Repellent Textiles and Clothing. Woodhead Publishing; 2018. pp. 233-265

[21] Rahmani-Azad M, Najafi A, Rahmani-Azad N, Khalaj G. Improvement of ZrB2 nanopowder synthesis by sol-gel method via zirconium alkoxide/boric acid precursors. Journal of Sol-Gel Science and Technology. 2022;**103**(1):87-96

[22] Enke D, Gläser R, Tallarek U. Sol-gel and porous glass-based silica monoliths with hierarchical pore structure for solid-liquid catalysis. Chemie Ingenieur Technik. 2016;**88**(11):1561-1585

[23] Jugović D, Uskoković D. A review of recent developments in the synthesis procedures of lithium iron phosphate powders. Journal of Power Sources. 2009;**190**(2):538-544

[24] Hench LL, West JK. The sol-gel process. Chemical Reviews. 1990;**90**(1): 33-72

[25] Tripathi VS, Kandimalla VB, Ju H. Preparation of ormosil and its applications in the immobilizing biomolecules. Sensors and Actuators B: Chemical. 2006;**114**(2):1071-1082

[26] Otitoju TA, Okoye PU, Chen G, Li Y, Okoye MO, Li S. Advanced ceramic components: Materials, fabrication, and applications. Journal of Industrial and Engineering Chemistry. 2020;**85**:34-65

[27] Ouyang Y, Bai L, Tian H, Li X, Yuan F. Recent progress of thermal conductive ploymer composites: Al2O3 fillers, properties and applications. Composites Part A: Applied Science and Manufacturing. 2022;**152**:106685

[28] Schubert U. Chemistry and fundamentals of the sol–gel process. In: The Sol-Gel Handbook. 2015. pp. 1-28

[29] Gaweł B, Gaweł K, Øye G. Sol-gel synthesis of non-silica monolithic materials. Materials. 2010;**3**(4): 2815-2833

[30] Landau MV. Sol–gel process. In: Handbook of Heterogeneous Catalysis: Online. 2008. pp. 119-160

[31] Wang C-C, Ying JY. Sol− gel synthesis and hydrothermal processing of anatase and rutile titania nanocrystals. Chemistry of Materials. 1999;**11**(11): 3113-3120

[32] Sakka S. History of the Sol-Gel chemistry and technology. In: Klein L, Aparicio M, Jitianu A, editors. Handbook

of Sol-Gel Science and Technology: Processing, Characterization and Applications. Cham: Springer International Publishing; 2018. pp. 3-29

[33] Bokov D, Turki Jalil A, Chupradit S, Suksatan W, Javed Ansari M, Shewael IH, et al. Nanomaterial by sol-gel method: Synthesis and application. Advances in Materials Science and Engineering. 2021; **2021**:5102014. DOI: 10.1155/2021/5102014

[34] Woo H, Reucroft PJ, Jacob RJ. Electrodeposition of organofunctional silanes and its influence on structural adhesive bonding. Journal of Adhesion Science and Technology. 1993;7(7): 681-697

[35] Shacham R, Avnir D, Mandler D. Electrodeposition of methylated sol-gel films on conducting surfaces. Advanced Materials. 1999;**11**(5):384-388

[36] Liu L, Mandler D. Electrochemical deposition of sol-gel Films. In: Klein L, Aparicio M, Jitianu A, editors. Handbook of Sol-Gel Science and Technology: Processing, Characterization and Applications. Cham: Springer International Publishing; 2018. pp. 531-568. DOI: 10.1007/978-3-319-32101-1_113

[37] Zhou B, Wu Y, Zheng H. Investigation of electrochemical assisted deposition of sol-gel silica films for long-lasting superhydrophobicity. Materials. 2023;**16**(4):1417

[38] Brinker CJ, Frye G, Hurd A, Ashley C. Fundamentals of sol-gel dip coating. Thin Solid Films. 1991;**201**(1): 97-108

[39] Faustini M, Louis B, Albouy PA, Kuemmel M, Grosso D. Preparation of sol–gel films by dip-coating in extreme conditions. The Journal of Physical Chemistry C. 2010;**114**(17):7637-7645. DOI: 10.1021/jp9114755

[40] Rio E, Boulogne F. Withdrawing a solid from a bath: How much liquid is coated? Advances in Colloid and Interface Science. 2017;**247**:100-114

[41] Puetz J, Aegerter MA. Dip coating technique. In: Aegerter MA, Mennig M, editors. Sol-gel Technologies for Glass Producers and Users. Boston, MA: Springer US; 2004. pp. 37-48

[42] Birnie DP. Spin coating technique. In: Aegerter MA, Mennig M, editors. Sol-Gel Technologies for Glass Producers and Users. Boston, MA: Springer US; 2004. pp. 49-55. DOI: 10.1007/978-0-387-88953-5_4

[43] Butt MA. Thin-film coating methods: A successful marriage of high-quality and cost-effectiveness - A brief exploration. Coatings. 2022;**12**(8):1115

[44] Mahadik SA, Mahadik D, Kavale M, Parale V, Wagh P, Barshilia HC, et al. Thermally stable and transparent superhydrophobic sol–gel coatings by spray method. Journal of Sol-Gel Science and Technology. 2012;**63**:580-586

[45] Li F, Lan X, Wang L, Kong X, Xu P, Tai Y, et al. An efficient photocatalyst coating strategy for intimately coupled photocatalysis and biodegradation (ICPB): Powder spraying method. Chemical Engineering Journal. 2020;**383**: 123092. DOI: 10.1016/j.cej.2019.123092

[46] Liu Y, Huang J, Feng X, Li H. Thermal-sprayed photocatalytic coatings for biocidal applications: A review. Journal of Thermal Spray Technology. 2021;**30**(1):1-24. DOI: 10.1007/s11666-020-01118-2

[47] Gardon M, Guilemany JM. Milestones in functional titanium dioxide thermal spray coatings: A review. Journal of Thermal Spray Technology. 2014;**23**(4): 577-595. DOI: 10.1007/s11666-014-0066-5

[48] Sáenz-Trevizo A, Hodge A. Nanomaterials by design: A review of nanoscale metallic multilayers. Nanotechnology. 2020;**31**(29):292002

[49] Izyumskaya N, Alivov Y-I, Cho S-J, Morkoç H, Lee H, Kang Y-S. Processing, structure, properties, and applications of PZT thin films. Critical Reviews in Solid State and Materials Sciences. 2007; **32**(3–4):111-202

[50] Rahmani E, Ahmadpour A, Zebarjad M. Enhancing the photocatalytic activity of TiO_2 nanocrystalline thin film by doping with SiO_2. Chemical Engineering Journal. 2011;**174**(2):709-713. DOI: 10.1016/j.cej.2011.09.073

[51] Monshi A, Foroughi MR, Monshi MR. Modified Scherrer equation to estimate more accurately nano-crystallite size using XRD. World Journal of Nano Science and Engineering. 2012; **2**(3):154-160

[52] Monai M, Montini T, Fornasiero P. Brookite: Nothing new under the Sun? Catalysts. 2017; **7**(10):304

[53] Folkmann F, Gaarde C, Huus T, Kemp K. Proton induced X-ray emission as a tool for trace element analysis. Nuclear Instruments and Methods. 1974; **116**(3):487-499

[54] Goldstein JI, et al. Generation of X-rays in the SEM specimen. In: Scanning Electron Microscopy and X-ray Microanalysis. 3rd ed. Boston, MA: Springer US; 2003. pp. 271-296

[55] Andrade JD. X-ray photoelectron spectroscopy (XPS). In: Andrade JD, editor. Surface and Interfacial Aspects of Biomedical Polymers: Volume 1 Surface Chemistry and Physics. Boston, MA: Springer US; 1985. pp. 105-195

[56] Lee H-L, Flynn NT. X-ray photoelectron spectroscopy. In: Vij DR, editor. Handbook of Applied Solid State Spectroscopy. Springer; 2006. pp. 485-507

[57] Sun X, Tiwari D, Fermin DJ. High interfacial hole-transfer efficiency at $GaFeO_3$ thin film photoanodes. Advanced Energy Materials. 2020; **10**(45):2002784. DOI: 10.1002/aenm.202002784

[58] Giessibl FJ. Advances in atomic force microscopy. Reviews of Modern Physics. 2003;**75**(3):949

[59] Carpick RW, Salmeron M. Scratching the surface: Fundamental investigations of tribology with atomic force microscopy. Chemical Reviews. 1997;**97**(4):1163-1194

[60] Eaton P, West P. Atomic Force Microscopy. Oxford University Press; 2010

[61] Rahmani E, Rahmani M, Silab HR. TiO2:SiO2 thin film coated annular photoreactor for degradation of oily contamination from waste water. Journal of Water Process Engineering. 2020;**37**: 101374. DOI: 10.1016/j.jwpe.2020.101374

[62] Lauritsen JV, Vang RT, Besenbacher F. From atom-resolved scanning tunneling microscopy (STM) studies to the design of new catalysts. Catalysis Today. 2006;**111**(1–2):34-43

[63] Inkson BJ. Scanning electron microscopy (SEM) and transmission electron microscopy (TEM) for materials characterization. In: Hübschen G, et al., editors. Materials Characterization Using Nondestructive Evaluation (NDE) Methods. Woodhead Publishing; 2016. pp. 17-43

[64] Filippetto D, Musumeci P, Zolotorev M, Stupakov G. Maximum current density and beam brightness achievable by laser-driven electron sources. Physical Review Special Topics-Accelerators and Beams. 2014;**17**(2): 024201

[65] Ul-Hamid A. Components of the SEM. In: A Beginners' Guide to Scanning Electron Microscopy. Cham: Springer International Publishing; 2018. pp. 15-76

[66] Franken LE, Grünewald K, Boekema EJ, Stuart MC. A technical introduction to transmission electron microscopy for soft-matter: Imaging, possibilities, choices, and technical developments. Small. 2020;**16**(14): 1906198

[67] Na HR, Lee HJ, Jeon JH, Kim H-J, Jerng S-K, Roy SB, et al. Vertical graphene on flexible substrate, overcoming limits of crack-based resistive strain sensors. npj Flexible Electronics. 2022;**6**(1):2

[68] Reimer L. Elements of a Transmission Electron Microscope, in Transmission Electron Microscopy: Physics of Image Formation and Microanalysis. Berlin, Heidelberg: Springer Berlin Heidelberg; 1989. pp. 86-135

[69] Petford-Long AK, Chiaramonti AN. Transmission electron microscopy of multilayer thin films. Annual Review of Materials Research. 2008;**38**:559-584

[70] Al-Agel FA, Mahmoud WE. Synthesis and characterization of highly stoichiometric AgInSe2 thin films via a sol-gel spin-coating technique. Journal of Applied Crystallography. 2012;**45**(5): 921-925. DOI: 10.1107/S00218898120 3511X

Chapter 2

Theoretical Analysis of the Solid State and Optical Characteristics of Calcium Sulphide [CaS] Thin Film

Emmanuel Ifeanyi Ugwu

Abstract

The analysis of the calcium sulphide thin film material which is one of the families of chalcogenide groups of thin film materials was carried out in this work using a theoretical approach for which the propagated wave through the medium of the thin film that is deposited on a glass substrate is considered to be a scalar wave in nature. The thin film material is sectioned into twos, first section is termed homogeneous reference dielectric constant, ε_{ref} where no thin film is deposited on the substrate and the second part is termed perturbed dielectric function, $\Delta\varepsilon_p(z)$ containing the deposited thin film on the glass substrate. These two terms were substituted on the defined scalar wave equation that was subsequently solved using the method of separation of variable which invariably utilized in the transformation of the equation into the second type of Volterra equation. On the other hand, Green's function approach was also introduced in order to arrive at the model equation that culminates in an expression showcasing the wave propagated through the thin film material medium. This was subsequently applied in the computation of waves, $\psi(z)$ that is propagating through the material medium for various wavelengths within the ultraviolet, visible, and near-infrared region of the electromagnetic wave spectrum for which the influence of the aforementioned dielectric constant and function were invoked. The computed values from this mechanism were in turn utilized in the analysis of the band gap, optical, and solid-state properties of the calcium sulphide (CaS) thin film materials.

Keywords: scalar wave, dielectric constant, analysis, calcium sulphide, thin film, wave propagation, solid state and optical properties

1. Introduction

The Chalcogenide family of thin films to which calcium Sulphide belong is one of the sulphide-based thin film that has a wide range of applications and based on that, a lot of researchers have shown so much interest in its study and as a result, both experimental and theoretical techniques are being utilized to get into deep analysis to unravel more uniqueness the thin film. Based on this various growth techniques have been utilized to develop the thin film including the CBT growth technique with an emphasis on the study of its optical and structural properties [1]. Theoretically, mathematical tools have been used for the analysis of thin films of similar types by making use of their various properties in conjunction with wave propagating through the film medium. The first of its type was the use of beam propagation technique whereby the dielectric properties were employed in studying and computing beam or field propagation through a medium with variation in small refractive index [2, 3]. The beam propagation method based on diagonalization of the Hermetician operator that generates the solution of the Helmholtz equation in media with real refractive indices [4], has been utilized in this study by some researchers while others had somehow used 2x2 propagation matrix formalism for finding the obliquely propagated electromagnetic fields in layered inhomogeneous un-axial structure which also involved bean propagation [5]. Structures such as optical fibers and optical wave guides in the presence of electro-optical perturbation have been well understood by the application of this method [6–8]. Although earlier before then, work had been going on veraciously on the study of wave propagation in a stratified media, plasma and ionosphere that gave a more clear picture of atmospherics behavior as regards wave propagating through its medium [6, 9, 10]. Van Roey in his work derived a general beam propagation relation in a number of specific cases along with the extensive simulation of wave propagation in a variety of material mediums.

Scientists have also looked at the propagation of electromagnetic field through a conducting surface [3] where the behavior of wave propagated through such material coupled with the influence of the dielectric function of the medium on such material coupled with the effect of variation of the refractive index on some species of the thin film had been analyzed as well using the same approach [11]. And a close look at the concept made it clear that recognition of the importance of the effect of the refractive index of the medium and dielectric function culminated in the reality of the creation of two velocity components that normally give rise to phase and group refractive indices as considered in the study of wave propagation [12–15]. Recently more complicated work had been embarked upon on the study of wave propagation through a modeled thin with dielectric perturbation in which WKB approximation in conjunction with numerical approach were used [16–18] along with beam propagation technique to unravel the mechanism of theoretical analysis of wave propagation through materials and based on that, lots of work had been veraciously carried out in term of the influence of dielectric constants and refractive index on beam propagating through materials [3, 5, 12, 17, 19] in other to ascertain their roles and efficacy on the use of beam propagation method and again the effect of the refractive index of the medium in the reality of the two velocity components that normally give rise to phase and group refractive indices as generally considered when it comes to the study of wave propagation because their influence on the propagated beam [13]. This had gone a long way to add and to reveal the efficacy of the theoretical approach in understanding the beam propagation mechanism in wave propagating through thin film material. This is seen to be achieved due to the flexibility of Green's function as a tool because this is what facilitated the use of the iterative process that will be involved in the computational technique in this work that has enabled one to embark on this theoretical frame work [18, 19].

2. Theoretical procedures

General wave equation

$$\nabla^2\psi(z) = {}^1\!/\!_{c^2}\frac{\partial^2\psi(z)}{\partial t^2} \quad (1)$$

$$\nabla^2\psi(z) + \omega^2\mu_o\varepsilon_o\psi(z) = 0 \quad (2)$$

from which we obtain the Helmholtz form of it using the separation of variable technique which is one of the methods being used for the solution of the wave propagating through a medium in which a dielectric function as defined in our model in **Figure 1** which shown in Eq. (3).

$$\varepsilon(z) = \varepsilon_{ref} + \Delta\varepsilon(z) \quad (3)$$

consisting of two parts is imposed on the wave equation. The dielectric function consists of a perturbed part, $\Delta\varepsilon(z)$ representing the part where the thin film is deposited and the reference section where no film is deposited, ε_{ref}. Substituting the dielectric function in Eq. (2), we obtain [18]

$$\nabla^2\psi(z) + \mu_0\varepsilon_0\omega^2\psi(z) = -V(z) \quad (4)$$

where

$$V(z) = \gamma^2\varepsilon(z) \quad (5)$$

Green's function technique is used to obtain an expression for a wave propagating through the film as.

$$\psi(z) = \int_0^z G(z,z')V(z')\psi(z') \quad (6)$$

The Integral

$$\psi(z,z') = I(z,z') = \int_o^{z'} G(z,z')V(z')\psi(z,z'dz' \quad (7)$$

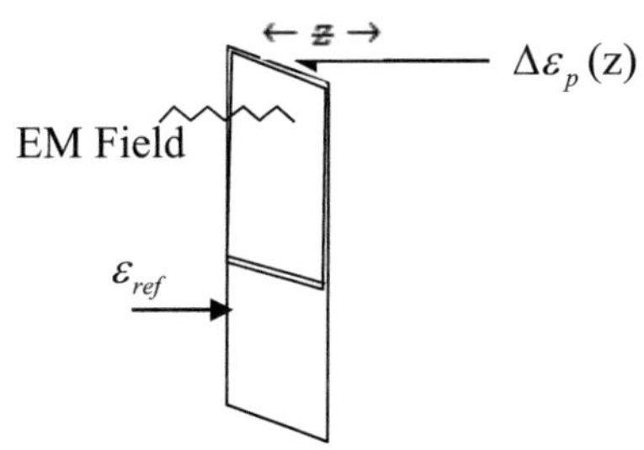

Figure 1.
Scalar wave impinging upon the substrate with reference medium ε_{ref} and the thin film medium $\Delta\varepsilon_p(z)$.

This when critically considered is a homogeneous Volterra equation of the second type with the kernel.

$$k(z,z') = G(z,z')V(z') \tag{8}$$

In this equation, the Neumann series method is not applicable to this type of equation; hence we apply Born approximation method that enables us to rewrite the equation as

$$\psi(z,z') \equiv I(z,z') = \int_o^{z'} \Phi(z,x)V(x)\psi(z,x)dx \tag{9}$$

where we put x in place of z'.

However, according to Born approximation procedure, we replace the unknown function $\psi(z,x)$ in the integral with a known function $\tilde{\Phi}(z,x)$ and use it to get an approximate solution of $\psi(z,z')$.

That is

$$\psi(z,z') = I(z,z') = \int_o^{z'} \Phi(z,x)V(x)\tilde{\Phi}(x) \tag{10}$$

The sign on $\Phi(z,x)$ is signified as an indication of the possible phase difference between the incoming wave and the outgoing wave. Thus neglecting the influence of the reflected wave in the system, we then use Green's function as in this case, we have.

$$G(z,z') = \Phi(z,z') = \frac{2}{z}\sum_{n=1}^{\infty} \frac{\sin n\pi z'/_z}{\gamma^2 - \frac{n^2\pi^2}{z^2}} \tag{11}$$

This results in the solution as given

$$\begin{aligned}\psi(z,z') &= I(z,z') \\ &= -\sum_{n=1}^{\infty} \frac{\gamma^2}{\gamma^2 - \frac{n^{2\pi^2}}{z^2}} \left[\frac{(\varepsilon' + cz')\cos\left(\frac{n\pi}{z} + \frac{2\pi}{\lambda}\right)z'}{\frac{n\pi}{z} + \frac{2\pi}{\lambda}} \right. \\ &\quad \left. + \frac{c}{\left(\frac{n\pi}{z} + \frac{2\pi}{\lambda}\right)^2} \sin\left(\left(\frac{n\pi}{z} + \frac{2\pi}{\lambda}\right)z'\right) - \left(\frac{\varepsilon' + cz'}{\frac{n\pi}{z} + \frac{2\pi}{\lambda}} \frac{c}{\left(\frac{n\pi}{z} + \frac{2\pi}{\lambda}\right)^2}\right)\right]\end{aligned} \tag{12}$$

$\psi(\lambda)$ vs.λ when z and z' are fixed.

In all use, we considered in the analysis the wavelength within λ = 250 nm to 1200 nm.

From the equation, it is obvious that ($\psi(z)$ tends to 0 as the dielectric constant $\varepsilon' \to \infty$,

Therefore, for fixed values of other parameters, the resultant solution can be approximated to any number of terms as may be required in relation to wave propagation terms.

This is used to obtain the absorption co-efficient using [20]

$$I = I_o \exp \alpha z$$
$$\alpha = {}^1\!/_z In\left(\frac{1}{T}\right) \tag{13}$$

known as Lamber-Beer- Bouguer law

The absorption coefficient is used in Eq. [21] to obtain the expression below [22].

$$(\alpha h\nu)^2 = A\left[h\nu - E_g\right] \tag{14}$$

In the case of dielectric, further deduction was involved since it is known that the refractive index and dielectric function which appears in both real and imaginary parts characterize the optical properties of any material because they are related to refractive index, n, and extinction coefficient, k.

$$\varepsilon_r = n^2 + k^2 \tag{15}$$

$$\varepsilon_i = 2nk \tag{16}$$

3. Results/discussion

Figure 1 represents a diagram showing the deposited thin film on the glass substrate which depicts a model of the thin film that is considered to represent a dielectric function with perturbed and the other part where there is no thin film we assume to be the termed reference section. This concept is mathematically represented in Eq. (3). And this equation on the other hand was used in conjunction with the general scalar wave equation to formulate a second-order differential equation in terms of dielectric function that was solved using the method of separation of variables and invariably applied to formulate Helmholtz equation as in Eq. (4). Eq. (3) was substituted in (2) in order to come up with an expression that signifies a propagating was through the thin film medium in terms of Green's function. The further deduction was made to harmonize the equations in agreement with Green's function which depicts a function that is in line with the homogeneous Volterra equation of the second type as in Eq. (7). The equation was invariably solved using the Born approximation approach that culminated in Eq. (10) that indicated the phase difference between the incident wave on the thin film and the outgoing wave of which the reflected part of the wave was ignored for the entire system in order to enable the analysis to be carried out in terms of transmitted waves as it propagates through the material thin film. The result of the solution is given in Eq. (12) which depicts $\psi(\lambda)$ as function of wavelength. The concept was finally made use of in the computation and deduction of absorption co-efficient as in Eq. (13) which formed the backbone of deduction as it concerns all the required optical and solid state properties of the material thin film as required to be analyzed in this work.

Secondly, this also was utilized in obtaining the bad gap as shown in **Figure 2** of the thin film based on the model that was presented in **Figure 1** which was presented separately in terms of all the wavelengths within the UV, Visible and near-infrared regions of electromagnetic wave spectra. The graph which depicts a plot of $(\alpha h\nu)^2$ as a function of wavelength within the three regions showcased the extrapolation that

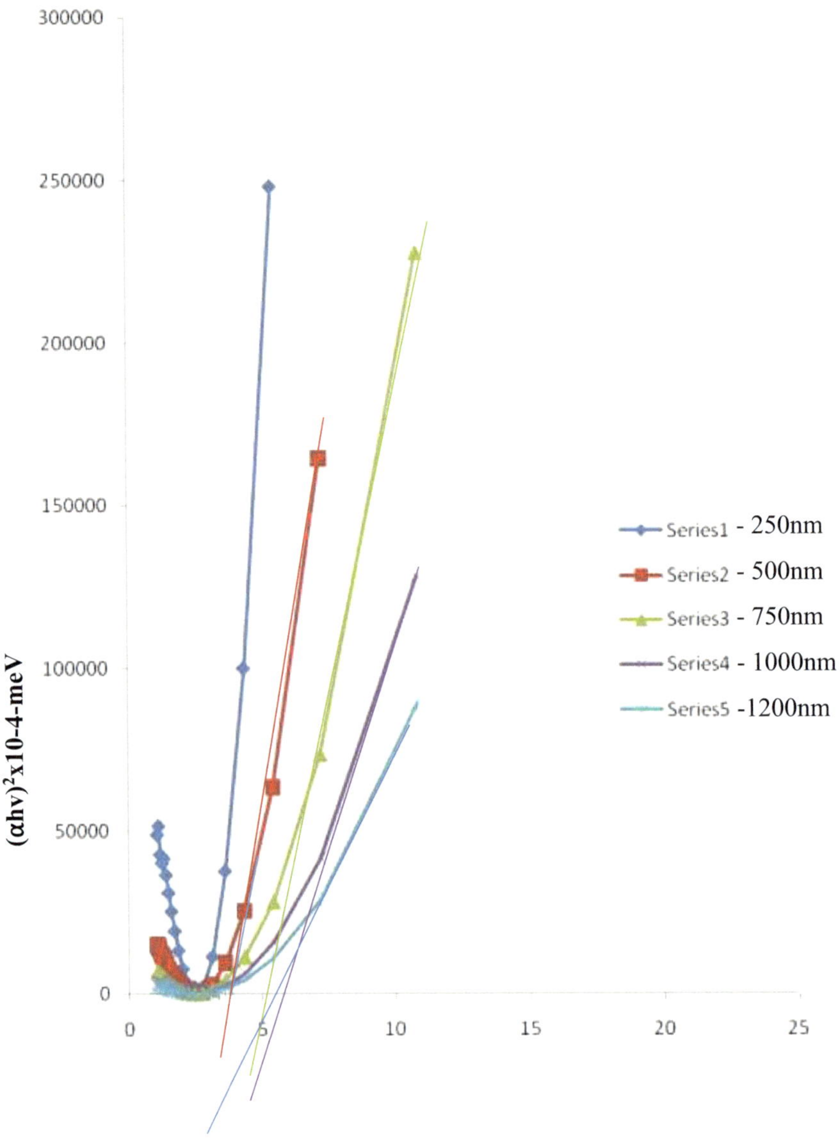

Figure 2.
A graph of $(\alpha h\nu)^2$ *as a function of wavelength for UV, visible and infrared of electromagnetic wave region (CaS) thin film.*

indicates the positions of their respective band gaps within the wavelength of the respective spectrum.

In determining the optical properties, the same process was carried out in determining the percentage transmittance, reflectance and absorbance as in **Figures 3–5**, which were also presented in terms of the wavelength spectrum of the considered region respectively. In the case of transmittance, percentage transmittance was used and it was discovered that within the visible and near-infrared it is zero with fluctuation up to a point at the UV region where it rose sharply while the reflectance appeared to be negative except at a particular point within the visible

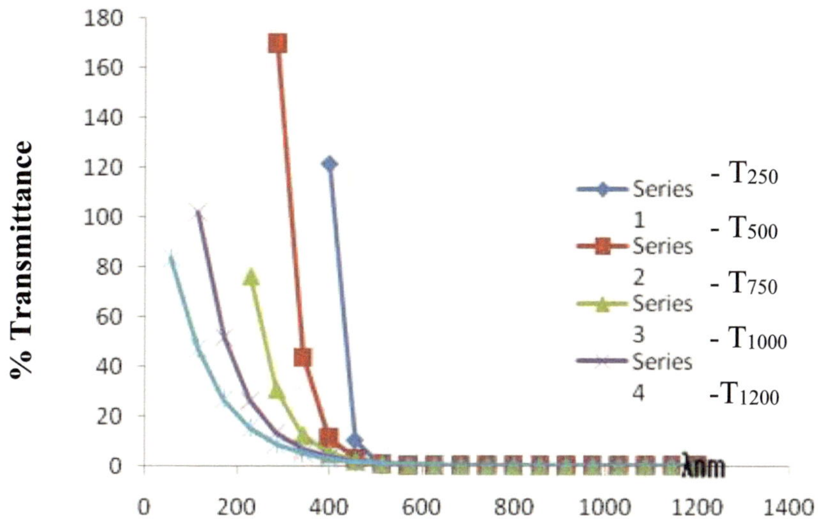

Figure 3.
Transmittance VS wavelength, λnm for UV, visible and infrared of electromagnetic wave region, calcium sulphide (CaS) thin film.

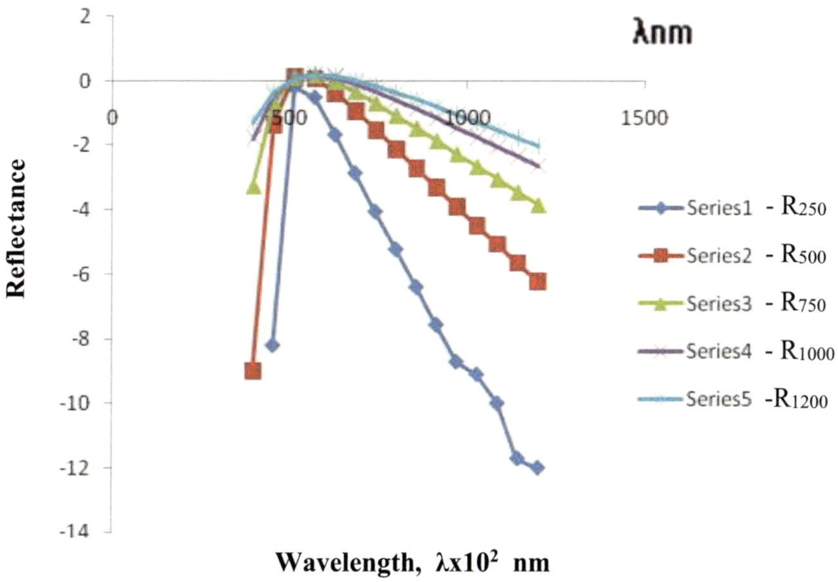

Figure 4.
Reflectance VS wavelength, λ for UV, visible and infrared of electromagnetic wave region; calcium sulphide (CaS) thin film.

where it grazes the axis at zero at the point on the visible region. From observation as depicted in the graph, the absorbance seems to be negative as well.

The dielectric constants of the film were obtained by considering the fundamental electron excitation spectrum of thin films is described by means of a frequency-dependent dielectric constant that is related to n and k as shown in Eqs. (15) and (16). However, the graph showing the plot dielectric constant for both real and imaginary

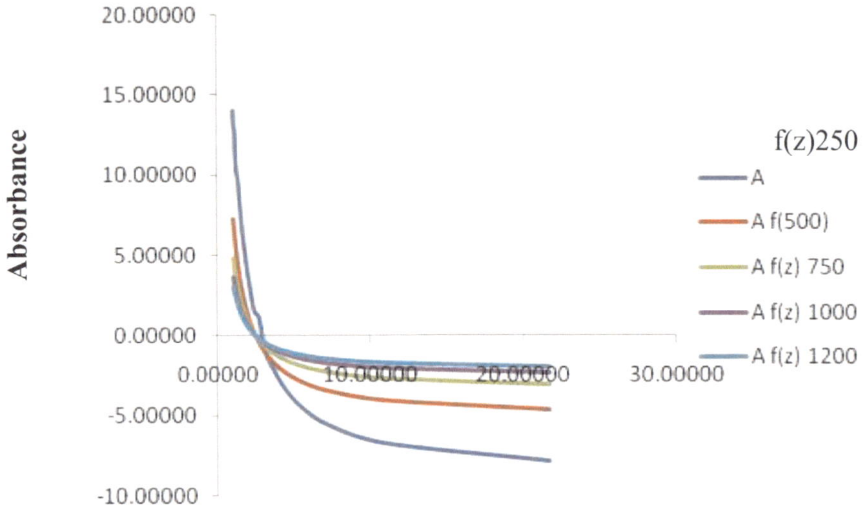

Figure 5.
Absorbance VS wavelength, λ_{nm} for UV, visible and infrared of electromagnetic wave region; calcium sulphide (CaS) thin film.

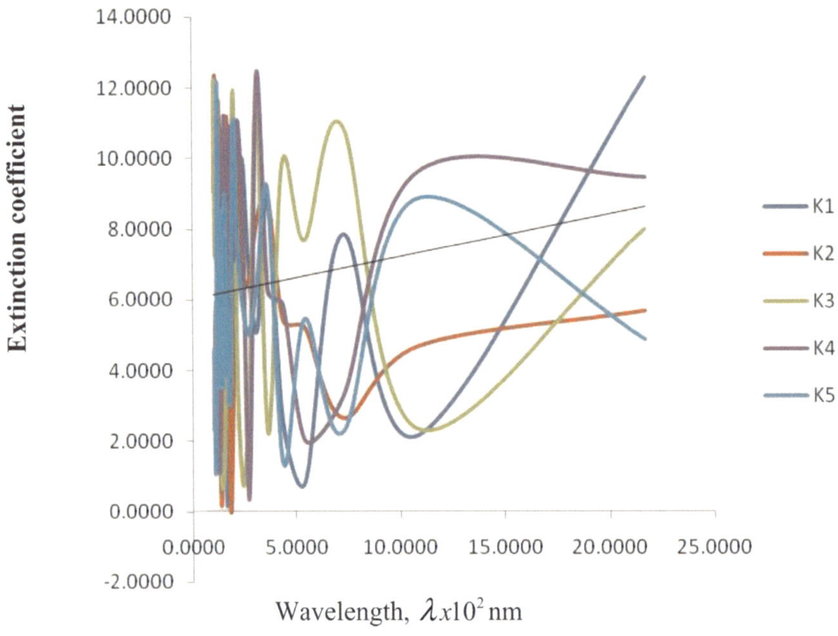

Figure 6.
Real dielectric constant VS wavelength, λ for UV, visible and infrared of electromagnetic wave region; calcium sulphide (CaS) thin film.

parts respectively were plotted as a function of wavelength as considered in this work and were shown in **Figures 6** and 7 while the extinction co-efficient is depicted in **Figure 8** was also plotted in the same manner.

However, it was generally observed that the behavior of the duo appeared to be irregular in their pattern.

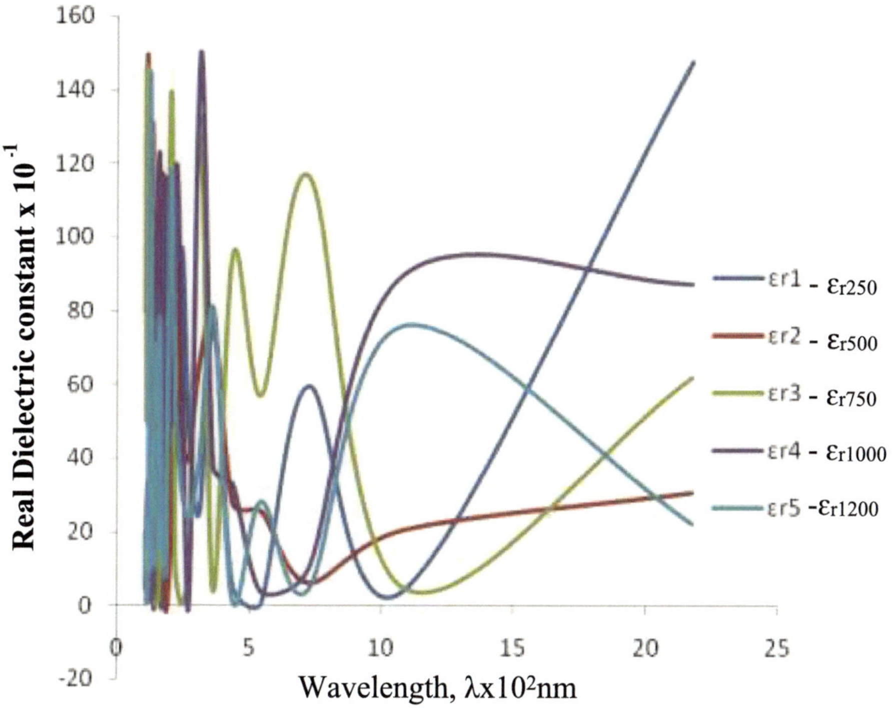

Figure 7.
Imaginary dielectric constant VS wavelength, λ for UV, visible and infrared of electromagnetic wave region; calcium sulphide (CaS) thin film.

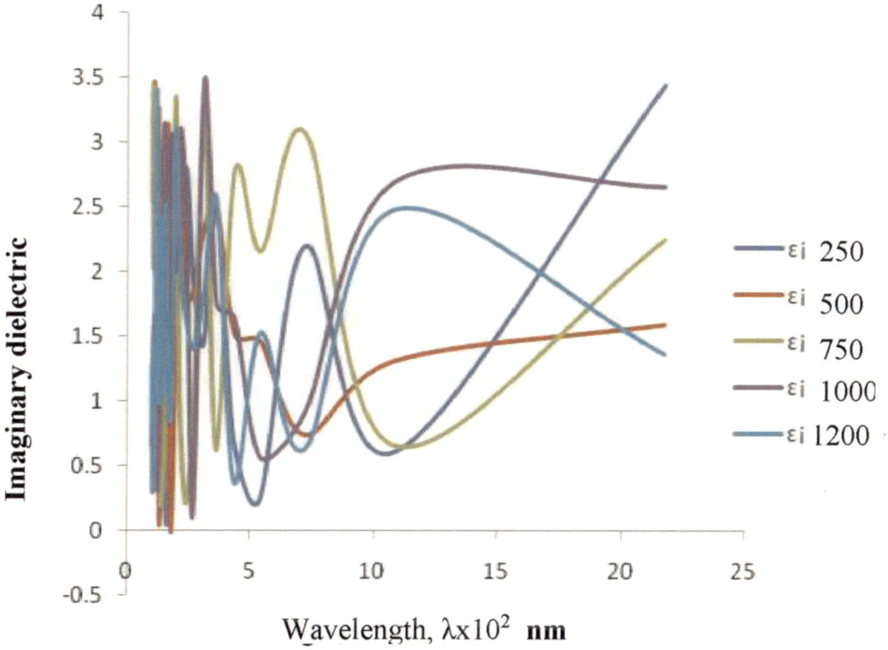

Figure 8.
Extinction coefficient constant VS wavelength for UV, visible and infrared of electromagnetic wave region; calcium sulphide (CaS) thin film.

Band gaps			
Thin film	**Computed band gaps (eV)**	**Average computed band gap (eV)**	**Experimental values of band gap (eV)**
CaS			
(i)	250 nm – 3.06		
(ii)	500 nm – 2.53		
(iii)	750 nm – 2.42	2.51	3.00–3.39
(iv)	1000 nm – 2.28		
(v)	1200 nm – 2.24		

Table 1.
Computed band gap as indicated by the extrapolated graphs.

Also, the energy band gap as deduced from the figure for various wavelengths have been shown in **Table 1** based on the considered regions of the EM wave.

The electromagnetic wave spectra propagated through the thin film material. As indicated, each wavelength has a unique band gap associated with it as shown in **Table 1** just as deduced from **Figure 2**. It is observed that the band gap for near-infrared is seen to be narrower while that of UV (i) is wider. However, the average of the duo is considered to be the actual band gap of the thin film that is 2.51 eV. At the initial stage, it was inferred from the literature based on the experimental result that the band gap of the thin film is within the range of 3.00–3.39 eV. Thus, appropriately it is reasonably considered that the computed band gap of CaS is narrower than the result from the experimental value though with just little margin of about $\sim$0.697 eV.

4. Conclusion

The analysis of the optical and solid state properties of CaS thin film has been carried out successfully using a general scalar wave equation that was made solvable by the use of Green's function technique and which finally led to the deduction of the propagation wave through the CaS thin film material. However, from the computed result and analysis in terms of the energy band gaps, it was discovered that there is a discrepancy between the values of the energy band gaps as observed in the table. This is the fact that the computed band gap is narrower than the one obtained experimentally However, the reason may not be farfetched from the assumptions and approximations that were involved during the mathematical deductions in formulating the governing equations that were used in computation which might, of course, affected the computed results of the band gaps, and perhaps the results of other graphs that showcased the optical properties as shown in graphs and also as recorded in the table some of which often appeared to be negative contrary to the experimental results in so many cases. From the computation and analysis, it was discovered that the computed band gap is narrower than the one obtained experimentally.

However, the unique feature of this work is that it has indicated that the energy band gap of any material can be studied in terms of the wavelength of the radiation propagating through the material since the wavelength of every region of the electromagnetic wave spectrum has a band gap associated with it.

Author details

Emmanuel Ifeanyi Ugwu
Department of Physics, Nigerian Army University, P.M.B, Biu, Nigeria

*Address all correspondence to: ugwuei2@gmail.com; ugwuei@yahoo.com; emmanuel.ifeanyi@naub.edu.ng

References

[1] Lokhande CD, Ennaoui A, Patil PS, Giersig M, Muller M, Diesner K, et al. A process and characterization of chemical bath deposited manganese sulphide. Thin Solid Films. 1998;**330**(2):70-75

[2] Feit MD, Fleck JA. Calculating dispersion in graded index multimode fibres by propagating beam method. Applied Optics Letters. 1979;**18**:2843-2851

[3] Feit MD, Fleck JA. Computation of mode properties in optical fibres wave guides by a propagating beam method. Applied Optics. 1980;**19**:1154-1164

[4] Ugwu EI, Uduh PC, Agbo GA. The effect of change in refractive index on wave propagation through (FeS_2). Journal of Applied Science. **7**(4):570-574

[5] Thylen L, Lee MC. Beam propagationmethod based on matrix diagonalization. Journal of Optical Society. 1992;**42**:146

[6] OngH L, Mayer RB. Electromagnetic wave propagation of polarized light in anisotropic media, application to liquid crystal. Journal of the Optical Society of America. 1983;**73**:167-176

[7] Van Roey JJ, Vander D, Lagasse PE. Beam propagation method: Analysis and assessment. Journal of the Optical Society of America. 1981;**71**:7

[8] Kim L, Gustafson TKG, Thylen L. An analysis of quantum confide structures using the beam propagation method. Applied Optical Letters. 1990;7:285-287

[9] Wait JR. Electromagnetic Wave in a Stratified Media. 2nd ed. Oxford, London: Pergamon Press; 1970

[10] Ginzbury VT. Electromagnetic Wave in Plasma. Oxford, London: Pergamon Press; 1967

[11] Ugwu EI, Uduh PC. Effect of the electrical conductivity of FeS_2 thin film on E.M wave propagation. JICOTECH Maiden Edition. 2005;**2005**:121-127

[12] Valanju PM, Walser RM, Valanju PA. Wave refraction in negative-index media; Always negative and very inhomogeneous. Physical Review Letters. 2002;**88**:187401

[13] Pentry JB, Smith DR. Commentary on 'wave refraction in negative- index media: Always positive and very inhomogeneous'. Physical Review Letters. 2003;**90**:029703

[14] Cox PA. The Electronic Structure and Chemistry of Solid. Oxford, London: Oxford University; 1978

[15] Ugwu EI. Theoretical study of field propagation through a nonhomogeneous thin film medium using Lippmann-Schwinger equation. The International Journal of Multiphysics. 2010;**4**(4):305-315

[16] Ong HL. 2x2 propagation matrices for electromagnetic wave propagating obliquely in layered inhomogeneous uni-axial media. Journal of the Optical Society of America. 1993;**10**(2):283-293

[17] Ugwu EI, Eke Vincent OC, Onyekachi E. Study of the impact of dielectric constant perturbation on electromagnetic wave propagation through material medium: MathCAD solution. Chemistry and Material Research. 2012;**2**(6)

[18] Martin JF, Alain D, Christian G. Alternative scheme of computing exactly

the total field propagating in dielectric structure of arbitrary shape. Journal of Optical Society America A. 1994;**1**(3): 1073-1080

[19] Ugwu EI. Optical and solid state properties of manganese sulphide thin film: Theoretical analysis. International Journal of Multiphysics. 2017;**11**(2): 137-150

[20] Ugwu EI. Comparative analysis of spectral properties of antimony selenide and copper sulphide thin film: Beam propagation computational approach. Advanced in Theoretical & Computational Physics. 2020;**3**(3):228

[21] Chopra KL, Kainthla DK. Physics of Thin Film. New York: Academic Press; 1982. pp. 169-235

[22] Nadeem MY. Optical properties of ZnS thin film. Turkish: The Journal of Physiology. 2000;**24**:651-659

Chapter 3

From Challenges to Solutions, Heteroepitaxy of GaAs-Based Materials on Si for Si Photonics

Junjie Yang, Huiwen Deng, Jae-Seong Park, Siming Chen, Mingchu Tang and Huiyun Liu

Abstract

Monolithic growth of III-V materials onto Si substrates is appealing for realizing practical on-chip light sources for Si-based photonic integrated circuits (PICs). Nevertheless, the material dissimilarities between III-V materials and Si substrates inevitably lead to the formation of crystalline defects, including antiphase domains (APBs), threading dislocations (TDs), and micro-cracks. These nontrivial defects lead to impaired device performance and must be suppressed to a sufficiently low value before propagating into the active region. In this chapter, we review current approaches to control the formation of defects and achieve high-quality GaAs monolithically grown on Si substrates. An APB-free GaAs on complementary-metal-oxide semiconductor (CMOS)-compatible Si (001) substrates grown by molecular beam epitaxy (MBE) only and a low TD density GaAs buffer layer with strained-layer superlattice (SLS) and asymmetric step-graded (ASG) InGaAs layers are demonstrated. Furthermore, recent advances in InAs/GaAs quantum dot (QD) lasers as efficient on-chip light sources grown on the patterned Si substrates for PICs are outlined.

Keywords: heteroepitaxy, III-V materials, Si, defects, integration

1. Introduction

Recently, InP and GaAs-based optical transceivers have been progressively replacing the traditional copper interconnect due to the unique properties of high transmission speed, larger bandwidth, and less cooling power required [1, 2]. Although great performances of III-V-based optoelectronic devices have been demonstrated, the cost and scalability limit the entry of their applications into the market of consumer electronics and massive production [3, 4]. In contrast, low-cost, high-bandwidth, and high-speed Si-based PICs and optoelectronic integrated circuits (OEIC) are ideal candidates to replace high-cost InP and GaAs-based PICs due to the large scalability and better thermal conductivity [3, 5–8].

However, as a critical component of PICs, a highly reliable and efficient Si-based laser is missing due to the indirect bandgap structure of Si and Ge bulk materials

[9, 10]. Fortunately, most III-V materials have superior optical properties and are ideal to be used as a laser gain medium, but a suitable integration method is needed to combine III-V materials and Si platforms [11–18]. As one of the most mature techniques, wafer bonding has been commercially used in the Si optical transceivers but left a questionable yield and cost [19–23]. Even though the integration method of direct epitaxy of III-V materials on the Si platform could cause many types of crystal defects, which leads to a substantial deterioration in the device performance, it has great potential owing to various advantages of large-scale, high yield, low-cost, and dense integration [24].

The crystal defects generated at the III-V/Si interface will trap carriers and produce extra heat to the devices by forming nonradiative recombination centers [25, 26]. Hence, a proper strategy to reduce and eliminate these crystal defects becomes the most critical condition to realize high-performance Si-based III-V optoelectronic devices. In this chapter, we will introduce the generation of different types of crystal defects in terms of different physical dimensions, followed by their corresponding solution and recently demonstrated results. After that, recent advanced works of monolithic integration of III-V QD lasers on the Si platform along with optical waveguide are discussed.

2. Direct epitaxy of III-V materials on Si

Although the direct epitaxy of III-V material on Si offers substantial benefits, issues streaming from large material dissimilarities between these two materials lead to the generation of nontrivial defects. For instance, the different polarity, large mismatch of lattice constant, and incompatible thermal expansion coefficient (CTE) result in the formation of APBs, TDs, and micro-cracks, respectively. Extensive endeavors have been dedicated to advancing the growth techniques in the last decades to tackle these three main challenges in III-V/Si heteroepitaxy and realize high-performance III-V lasers integrated into Si. In this section, the mechanisms of defect formation and strategies to suppress the defects will be discussed.

2.1 Antiphase boundaries

A planar defect called APB is formed during the heteroepitaxy of polar III-V materials on nonpolar Si (001) substrates. Si (001) vicinal surface with a small offcut angle ($<1°$) exhibits terraces of alternating 1×2 and 2×1 dimerization, which are separated by Si single-atomic-height (S) steps [27–30]. These Si S steps are classified into two groups, that is, S_a and S_b, depending on the dimer orientation in the upper terrace [31]. The schematic diagram of alternating S steps on the Si (001) surface is shown in **Figure 1**, where the S_a steps are straight, and the S_b steps are meandering due to thermal fluctuation [28]. The terrace width between the adjacent S_a and S_b steps is defined as L. The relationship between terrace width L and the offcut angle of the Si substrate θ is defined as

$$a = L \times \tan\theta \qquad (1)$$

where a is the height of a single Si S step, corresponding to 0.136 nm.

In most zinc-blend structures, for example, III-As and III-P materials, different atoms occupy the two face-centered-cubic (FCC) sublattices. By contrast, identical

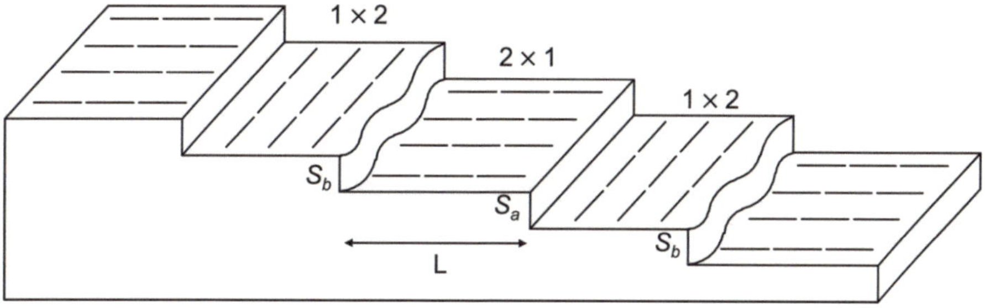

Figure 1.
Schematic diagram of Si (001) surface with S steps. Reprinted from reference [32] ©2021 the author under CC BY.

atoms occupy FCC sublattices in the diamond crystal structure of Si [33]. During the heteroepitaxy of III-V materials on Si (001) substrates, the orthogonal Si dimers in adjacent terraces lead to the formation of two domains with opposite sublattice allocation, that is, antiphase domains (APDs). The interface between two APDs is an APB consisting of homopolar bonds (Ga—Ga or As—As bonds) and is considered as electrically charged planar defect [34]. APBs propagate within the epilayers and act as nonradiative recombination centers and electrical leakage paths, severely degrading the optoelectronic properties of devices due to their relatively large area [34, 35]. In addition, the elastic strain associated with APBs will distort the crystal lattice and deform the DFLs. As a result, a high TD density (TDD) will be observed in the active region [32].

The presence of large-scale APBs can be characterized by using an atomic force microscope (AFM), electron scanning microscope (SEM), or transmission electron microscope (TEM). **Figure 2(a)** shows a typical AFM image of GaAs grown on Si (001) substrate with dense APBs, illustrated as curved boundaries. In practice, APBs emerge at the edge of Si *S* steps, and most of them propagate through {110} planes to the surface at low growth temperature as {110} APBs exhibit the lowest formation energy when compared with {112} and {111} APBs in both GaAs and GaP [36]. The kink of APBs into higher index planes, for instance, {111} and {112} planes, depends on the growth temperature of the epilayer. This process is crucial for the self-annihilation of APBs at the intersection [36, 37]. In stark contrast, the Si dimers are in the same orientation on the double-atomic-height (*D*) stepped Si surface. As a consequence, the nucleation of APBs is suppressed during the growth of III-V materials on Si *D* steps [25]. The APB nucleation and propagation under different circumstances are summarized in **Figure 2(b)**.

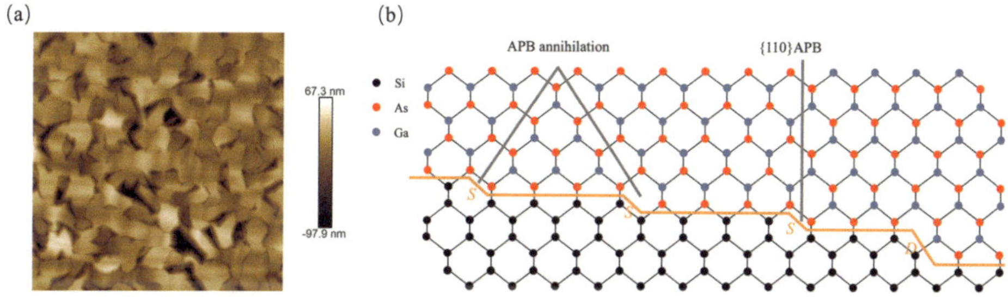

Figure 2.
(a) Top view AFM image shows dense APBs on the GaAs surface. (b) Schematic diagram of APB nucleation, propagation, and annihilation on Si steps.

A classic and common solution for suppressing APBs is to implement 4°–6° offcut Si substrates titled toward [110] direction, which preferentially forms *D* steps-dominated Si surface and thus inhibits the nucleation of APBs [28]. However, in order to be compatible with well-established CMOS processing technology, nominal Si (001) substrate with a misorientation of lower than 0.5° is required [38]. In the past decades, many techniques have been developed to achieve heteroepitaxy of APB-free III-V materials on on-axis Si (001) substrates, which will be introduced in the following contents.

2.1.1 Selective area growth

Selective area growth (SAG) allows III-V heteroepitaxy on the prepatterned Si substrates and attracts intensive scientific interest as it provides efficient defect reduction, attributed to epitaxy necking effects and aspect ratio trapping (ART) [39–41]. SAG of III-V materials on the narrow trenches with patterned vertical dielectric sidewalls (normally SiO_2) ensures sufficient TD trapping if a high aspect ratio (AR) is defined. The AR is defined as the ratio of trench height h and width w:

$$AR = \frac{h}{w} \tag{2}$$

During the epitaxy of mismatched III-V on Si substrates, the misfit dislocation (MD) inevitably formed at the interface due to strain relaxation. Glissile 60° MDs tend to form segments that thread up as TDs, which propagate freely on the {111} planes in <110> directions and move upwards to the epilayer surface [42, 43]. As for the TDs propagating on the {111} planes perpendicular or parallel to the trench orientation, TDs will eventually hit the vertical dielectric sidewalls and are trapped if sufficient AR is applied, as illustrated in **Figure 3(a)** case (1) and (2). Since {111} planes form an incident angle of 54.7° with [1$\bar{1}$0] direction, a minimum AR of 1.41 is required to terminate the TDs within the trench. Unlike the TDs, however, the trench only traps the planer defects lying on the {111} planes parallel to the trench orientation, as indicated by case (3) and (4) in **Figure 3**. Furthermore, V-grooved Si (001) substrates with {111} facets, formed by using wet etching, were developed to prohibit the formation of APBs even with the presence of *S* steps [25, 44], as shown in **Figure 3(b)**.

Through V-grooved Si {111} surfaces via an ART process, Li et al. successfully demonstrated an APB-free GaAs-on-V-grooved Si (GoVS) template [45]. The Si (001) substrate patterned with [110] direction SiO_2 strips was etched by KOH solution to form V-grooved Si {111} facets as the etching rate of Si being the lowest in the (111) plane

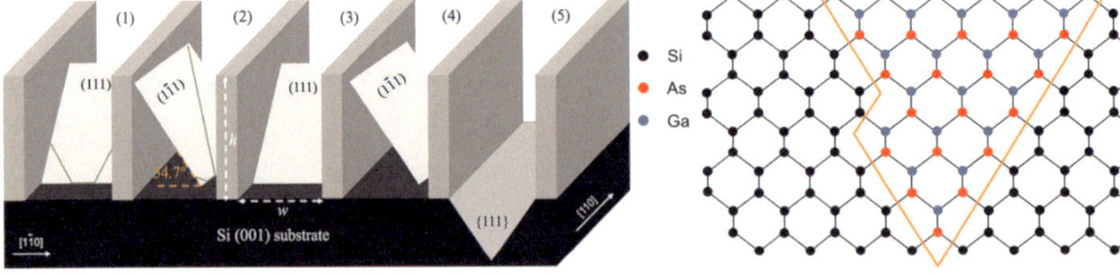

Figure 3.
(a) Schematic diagram showing TD and PD propagation within the narrow trenches with patterned vertical dielectric sidewalls. The APB nucleation is prohibited during the growth of III-V on V-grooved Si (001) substrate with {111} facets (b) even on a Si (111) S step.

[44]. SAG of GaAs nanowires was performed using a metalorganic chemical vapor deposition (MOCVD) system with a two-step growth method. **Figure 4(a)** shows the initial growth of GaAs on the V-grooved substrate, and planar GaAs nanowire is observed without APBs, thanks to the Si {111} facets. Interestingly, the unique "tiara"-like shape formed by Si undercut blocks the propagation of stacking faults through {111} planes, as shown in **Figure 4(b)**. After removing the SiO_2 strips by a buffered oxide etch, coalesced GaAs thin film was grown to finish the template. High-quality APB-free GaAs thin film was obtained after 300 nm GaAs overgrowth, as shown in **Figure 4(c)**.

In addition, Wei et al. proposed a novel way of forming Si {111} surfaces by homoepitaxy of Si on the U-shaped patterned Si (001) substrate. The U-shaped pattern along [110] direction is formed with a period of 360 nm, ridge width of 400 nm, and depth of around 500 nm by the deep ultraviolet photolithography (DUV) and subsequent dry etching, as shown in **Figure 5(a)** [46]. The U-shaped patterns will then be dipped in a diluted hydrofluoric acid to form a hydrogen-terminated surface and transferred into an IV molecular beam epitaxy (MBE) chamber for the deoxidation and subsequent homoepitaxy of 500 nm Si to form {111} facets, as shown in **Figure 5(b)**. Further deposition of Si to 550 nm leads to the merging of Si ridges and finally forms evolvement of (111)-faceted-sawtooth surface, which promotes the APB-free GaAs in the subsequent growth.

2.1.2 MOCVD/MOVPE grown APB-free GaAs/Si (001)

Although the pioneering works on SAG growth of APB-free III-V/Si (001) have been proved promising, the sophisticated processing and patterning of the Si surface

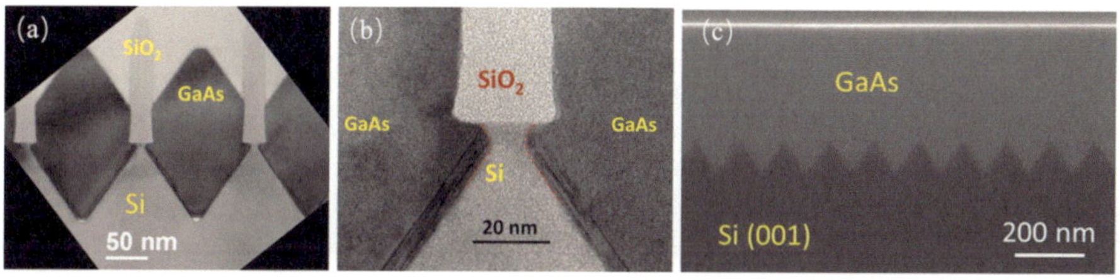

Figure 4.
(a) Cross-sectional TEM image showing the growth of GaAs on the V-grooved Si (001) substrates. (b) Defects are trapped by the "tiara"-like shape formed by Si undercut. (c)cross-sectional SEM of grown GaAs in the GoVS template. Reprinted with permission from [45] ©2015 AIP publishing.

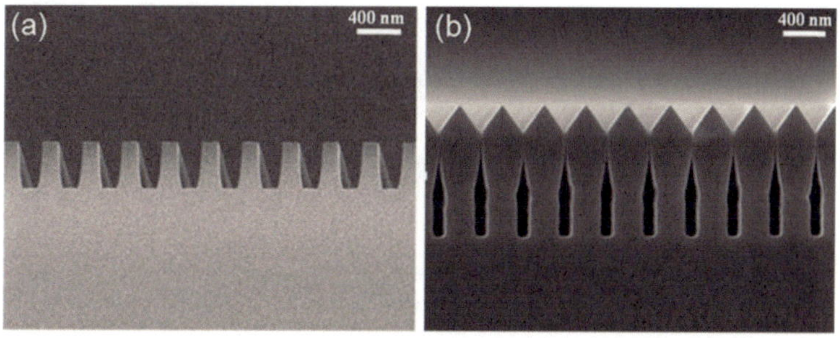

Figure 5.
(a) Cross-sectional SEM image showing a U-shaped patterned Si (001) substrate. (b) Si {111} facets formed by Si homoepitaxy. Reprinted with permission from [46] ©2015 AIP publishing.

are costly and time-consuming. Regarding the direct growth of III-V materials on planer Si (001) substrates, forming the *D* steps-dominated Si surface is the most straightforward idea to solve the APB issue. Thus, it has also been widely investigated. The Si *D* steps can be formed by high-temperature annealing of Si (001) substrates with proper hydrogen chemical potential, attributed to the preferential and selective etching of S_b steps by hydrogen [34]. This process is strongly related to the width of neighboring Si terraces. According to Eq. (1), a slightly large offcut angle of >0.1° is desired to erase S_b steps to a sufficiently low value and form the *D* steps-dominated Si surface.

Volz *et al.* achieved APB-free GaP on Si (001) substrate with an offcut angle of ~0.12° using metalorganic vapor phase epitaxy (MOVPE) [47–49]. A 500 nm Si buffer layer was first deposited, followed by postgrowth hydrogen annealing with the pressure of 950 mbar for 10 min at 975°C to obtain a *D* steps-dominated Si surface with an average terrace distance of ~120 nm. Nevertheless, *S* steps still appeared as a triangle-shaped form and existed between two neighboring *D* steps, covering ~15% of the surface area [37]. During the epitaxy of GaP on the Si buffer layer, APBs exist because of imperfect Si *D* steps. The distribution of APBs that resembles the underlying Si *S* steps is illustrated in **Figure 6(a)**, where the triangle-shaped *S* steps lead to the formation of APBs in two orthogonal directions [110] and [$\bar{1}$10]. The kinking and self-annihilation of APBs through energy favorable {112} is observed in the direction [110], which is perpendicular to the Si step orientation [36, 48]. The typical basal width of 180 nm for the *S* terrace yields a maximum height of 65 nm for APBs to be fully annihilated in GaP, as presented in **Figure 6(b)**. Since the triangle-shaped *S* steps become narrow along [110] direction, the decrease in basal length results in faster annihilation of APBs, which is confirmed by the TEM measurements in **Figure 6(c)**.

Based on this technique, a GaAs/GaP/Si (001) is developed as one of the most commercially successful templates for developing Si-based on-chip light sources. Many remarkable results have been reported based on this platform [50–52].

In contrast, Alcotte et al. selected a Si (001) substrate with a slightly larger miscut angle of 0.15° toward [110] direction to further enhance the etching of S_b steps. They demonstrated an APB-free GaAs grown on a Si (001) substrate by MOCVD without intermediate Si buffer layers [35]. Prior to growth, a Si wafer was first deoxidized in a SiConi™ chamber using an NF_3/NH_3 plasma. High-temperature hydrogen annealing at 850 to 950°C was then carried out to form Si *D* steps, as indicated in **Figure 7(b)**. A surface of GaAs grown on the un-optimized Si substrate is shown in **Figure 7(a)**,

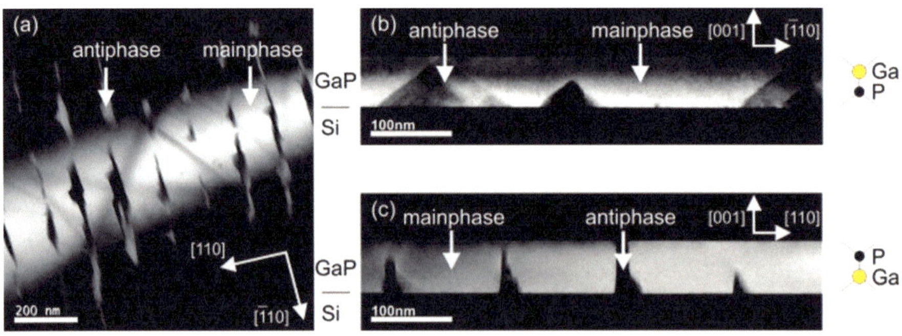

Figure 6.
(a) TEM plane view of GaP grown on pretreated Si buffer layer. Cross-sectional TEM measurement in (b) [110] and (c) [−110] cross-sections showing anisotropic APDs. Reprinted with permission from [48] ©2011 AIP publishing.

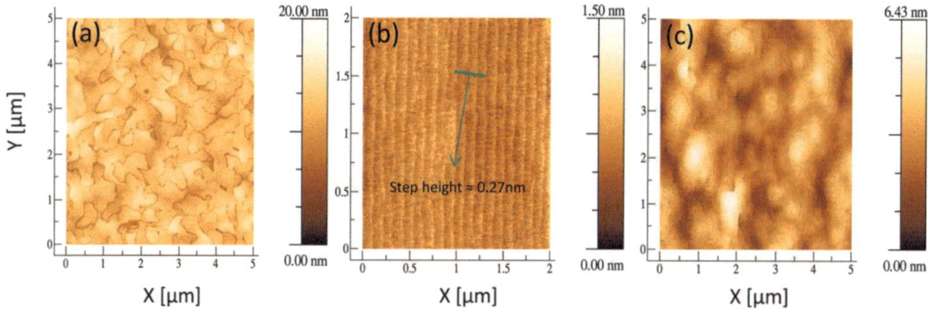

Figure 7.
(a) AFM image of GaAs grown on an un-optimized Si substrate showing high density of APBs. (b) Si D steps formed after annealing a Si substrate under optimized conditions. (c) AFM image of a 150 nm APB-free GaAs layer grown on Si (001). Reprinted from reference [35] ©2016 the authors under CC BY.

where a high density of APBs is visible. By utilizing an optimized Si substrate with the dominated Si *D* steps, a 150 nm APB-free GaAs epilayer was obtained with a low surface roughness of 0.8 nm for a 5×5 μm^2 AFM scan, as presented in **Figure 7(c)**.

2.1.3 MBE grown APB-free GaAs/Si (001)

Indeed, the above-mentioned APB-free III-V templates grown by MOCVD/MOVPE have achieved great success in commercialization. Nevertheless, the requirement of a hydrogen source is unsuitable for migrating such methods into MBE systems. The MBE system has a unique advantage in obtaining high-quality QDs [53], which are insensitive to defects and have been regarded as one of the most promising gain media for high-performance Si-based on-chip laser sources [54]. Developing an APB-free III-V layer by a single system simplifies the growth process and is economical in the long term.

This need was first satisfied by Kwoen et al. who have successfully grown APB-free III-V lasers on on-axis Si (001) by MBE using a high-temperature $Al_{0.3}Ga_{0.7}As$ nucleation layer (NL). In the study, four samples with identical structures except for the composition of Al in the first 40 nm $Al_xGa_{1-x}As$ NL were grown and compared by SEM and photoluminescence (PL). It was concluded that $Al_{0.3}Ga_{0.7}As$ NL promoted self-annihilation of APBs and delivered the best GaAs quality. Based on this platform, InAs/GaAs-based QD lasers were developed with high operating temperature [55, 56].

Recently, Li and Yang et al. proposed a new method of using periodic Si *S* steps to redistribute the APB nucleation and promote the APB annihilation during optimized GaAs overlayer growth by a dual-MBE system [57, 58]. In their study, on-axis Si (100) substrates with unselected miscut angles of $0.15 \pm 0.1°$ toward <110> were deoxidized at 1200°C for 30 min in the group-IV MBE. A 100 nm Si buffer layer was first grown at 850°C by using a Si e-beam source. This was followed by five iterations of 20 nm thin Si grown at 850°C and annealed at 1200°C to reconstruct the surface and form periodic Si steps. The wafer was then transferred into the III-V chamber for subsequent growth. The schematic image showing the APB-free GaAs buffer layer structure is illustrated in **Figure 8(a)**, which started with a low temperature (LT) $Al_{0.4}Ga_{0.6}As$ NL grown at 330°C, and the growth rate was 0.1 monolayers per second (MLs^{-1}). A temperature ramping step with a ramp rate of 10°C Min^{-1} was applied afterward. At the same time of increasing temperature, GaAs was deposited

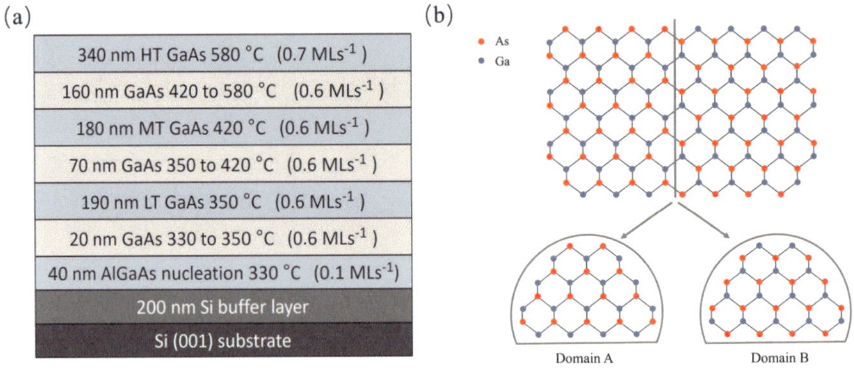

Figure 8.
(a) Schematic image of the APB-free GaAs buffer layer structure. (b) APB-MTE formed during high-temperature annealing. Reprinted from [58] © 2022 the authors, under CC BY.

simultaneously at a rate of 0.6 MLs^{-1}. This growth-during-ramp method was also applied in the following temperature ramping steps. Following the $Al_{0.4}Ga_{0.6}As$ NL, a three-step GaAs growth technique was implemented, consisting of 190 nm LT, 180 nm mid-temperature (MT), and 340 nm high-temperature (HT) GaAs grown at 350, 420, and 580°C respectively. The temperature ramping step was inserted between these GaAs layers, and the total GaAs buffer layer thickness was 1 μm.

In contrast to the growth parameter proposed by Kwoen et al. [59], the NL was grown at LT of 330°C in this study to avoid the formation of APB-modified thermodynamic equilibrium (APB-MTE). In APB-MTE, a (110) APB tends to enlarge, resulting in the formation of two APDs with opposite polarity, as shown in **Figure 8(b)** [58, 60]. Most recently, this growth strategy has been proved efficient by Gilbert et al. as it maintains the terrace-driven nature of APBs in initial nucleation rather than nucleation-driven, leading to controllable APB burying in the GaAs overgrowth [61]. Besides, the growth-during-ramping method helps to elongate the {110} APBs while preventing the APB-MTE. In addition, a suitable high growth temperature aids in the reconfiguration of the APBs into higher index planes, ultimately promoting the annihilation of APBs.

To examine the impact of the Si buffer layer on APB annihilation, a comparative analysis of the surface morphologies was conducted between Si substrates with and without an annealed Si buffer layer, as shown in **Figure 9**. For the deoxidized Si surface, the random distribution of Si atomic steps is observed in **Figure 9(a)**. These

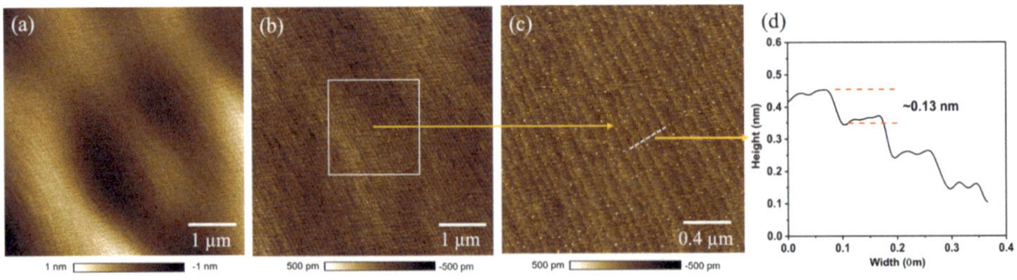

Figure 9.
5 ×5 μm^2 AFM image of (a) a deoxidized Si substrate and (b) a surface-reconstructed Si buffer layer. (c) 2 ×2 μm^2 AFM image of the Si buffer layer with periodic S steps. (d) Height measurement of Si steps, showing Si surface is mainly single-stepped. Reprinted from [57] © 2020 the authors, under CC BY.

undulating steps arise from the interaction between distinct stress domains on the Si surface, leading to the reduction in the overall elastic energy of the Si surface at a small offcut angle [62, 63]. In stark contrast, **Figure 9(b)** and **(c)** present the periodic *S* steps with a step height of ~0.13 nm, demonstrating only Si *S* steps instead of *D* presented on the Si surface after HT annealing [64, 65].

Two samples (sample A without Si buffer layer and sample B with Si buffer layer) with identical GaAs growth methods were grown and compared to test the impact of the Si *S* steps on APB annihilation. The cross-sectional TEM measurements are taken with a viewing direction of [$\bar{1}$10] for **Figure 10(a)–(d)** and [110] for (e) and (f). As shown in **Figure 10(a)–(d)**, APBs nucleate and propagate through the energy-favored {110} planes during LT GaAs growth. The APB propagation plane is configured to higher index planes in the high-temperature growth region, contributing to APB self-annihilation. The twisted patterns that demonstrate randomly distributed APB nucleation during GaAs/Si (001) (Sample A) are observed in **Figure 10(a)**. In contrast, periodic APBs occur when GaAs are grown on the Si buffer layer (Sample B), as illustrated in **Figure 10(c)**.

In both samples, APBs tend to intersect and annihilate with each other within the high-temperature growth region. However, in **Figure 10(a)**, the randomly distributed APBs shown in sample A propagate randomly within the GaAs, making them extremely difficult to eradicate effectively. The remaining APBs thus penetrate through the whole structure, as displayed in **Figure 10(b)**. In contrast, the well-organized APBs that nucleate on (S_a + S_b) resemble the underlying *S* steps and are closely spaced. The high growth temperature applied afterward sufficiently promotes the complete destruction of APBs within ~500 nm, as shown in **Figure 10(d)**.

Furthermore, the APBs that penetrate through the whole GaAs buffer layer in sample A are noticed from [110] viewing direction, as indicated in **Figure 10(e)**. By contrast, since the Si buffer layer is populated by *S* steps in [110] direction, the APBs that nucleate on these *S* steps resemble the step orientation, leaving no APB observed in [110] viewing direction, as shown in **Figure 10(f)**. This observation differs from the aforementioned APB-free GaP/Si growth, where triangle-island-shaped S_b steps

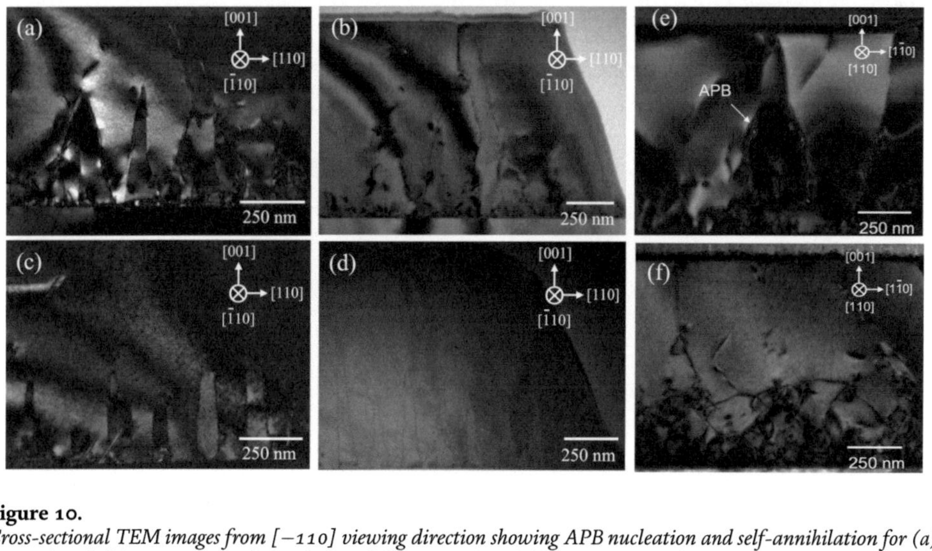

Figure 10.
Cross-sectional TEM images from [−110] viewing direction showing APB nucleation and self-annihilation for (a) regional and (b) 1 μ m range of sample a and (c) regional and (d) 1 μ m range of sample B. TEM images from [110] viewing direction for (e) sample a and (f) sample B. reprinted from [32] ©2021 the author under CC BY.

are left near the edge of the D steps. During subsequent growth of GaP on these S triangle islands, the APBs that resemble the underlying Si steps can be found in two orthogonal directions, that is, [110] and [$\bar{1}$10] [48, 49, 66].

The annihilation of APBs at HT is attributed to the difference in GaAs growth rate of the two domains. During the deposition of GaAs on Si, an Arsenic (As) prelayer is adopted to avoid Ga etching, and As resembles the underlying Si dimer orientation. Since Ga atoms diffuse mainly along the As dimer direction, GaAs deposited on the upper terrace of S_a are more likely to grow along the [110] direction (main phase), and GaAs deposited on the upper terrace of S_b will grow along [1$\bar{1}$0] direction (antiphase), as indicated in **Figure 11(a)** [67, 68]. The main-phase GaAs grow faster than antiphase GaAs in the [110] direction, forcing neighboring APBs to intersect toward each other during temperature increases. As a result, the annihilation of terrace-driven APDs is facilitated, as shown in **Figure 11(a)**.

Interestingly, it has been reported that {110} APBs help to reduce TDs during GaAs overgrowth [58]. {110} APBs trap {111} TDs to climb along it and promote the termination of TDs with opposite Burger vector signs. In addition, the trapped TDs can glide through other {111} planes and might be captured again by other {110} APBs. This process will recur until TDs are terminated or move beyond the APBs. Consequently, the TDD level reaches 8×10^8 cm^{-2} for GaAs grown on Si (001) substrate with periodic {110} APBs, which is half of GaAs grown on Si offcut substrate with identical growth structure. This result reveals the probability of controlling both APBs and TDs simultaneously and achieving a high-quality APB-free GaAs/Si (001) template by optimizing the GaAs growth technique in the future.

2.2 Dislocations

The second issue that hinders the direct epitaxy of III-V materials on Si is the formation of dislocations. Most III-V materials, except for GaP and aluminum phosphide (AlP), have a large lattice mismatch with Si. During the III-V-on-Si mismatched heteroepitaxy, the strain energy accumulated inside the epilayer is proportional to the epilayer thickness. Once the thickness of the strained layer exceeds a certain value, the so-called critical thickness, MDs, are formed at the interface to relax the accumulated strain. These MDs introduce missing or dangling bonds along the mismatched

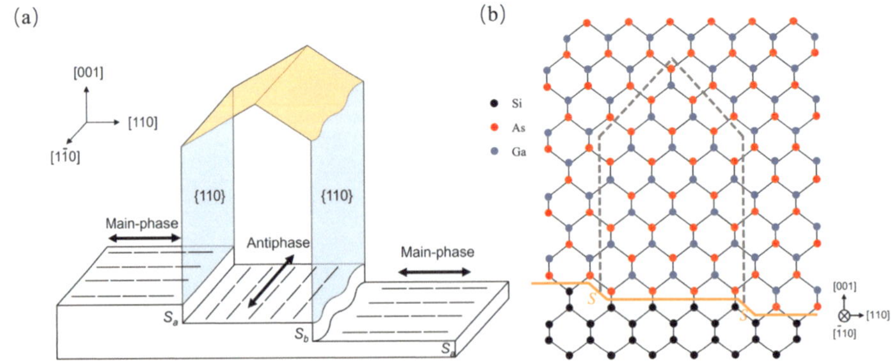

Figure 11.
(a) Schematic diagram of terrace-driven APB burying method. The area within the closed-loop APBs is considered as an antiphase, while the outside is the main phase. (b) Antiphase GaAs are buried by main-phase GaAs during the growth. Reprinted from [58]© 2022 the authors, under CC BY.

interface. The commonly occurred MDs can be classified into two types: (1) edge MDs with Burgers vector lying in the interface (001) plane and perpendicular to the line direction. This type of dislocation mainly originates from Si steps and is "sessile". (2) 60° MDs with Burgers vector of 60° to the dislocation line and 45° to the substrate, and this type of dislocation is termed "glissile". Since dislocations are one-dimensional defects and cannot terminate within a crystal, 60° MDs will move toward the edge of the crystal or form segments that thread up as TDs, which propagate freely along the {111} planes and penetrate through whole epi-layers. The TEM image of MD and TD are illustrated in **Figure 12**. TDs introduce deep states and act as nonrecombination centers for carriers, which leads to short carrier lifetime, low photon emission efficiency, and impaired device performance [43].

Several strategies were developed to control TDD in the past decades, aiming to reduce TDD to a low value of $\sim 10^{-6}$ and 10^{-5} cm^{-2}, which is close to the TDD on the native substrate. For instance, inserting a Ge intermediate or SiGe-graded buffer layer effectively bridges the lattice mismatch between Si substrates and III-V [69]. Besides, the three-step GaAs growth method is commonly utilized to control TDD within the GaAs buffer layer for the GaAs-on-Si system. A thin AlAs nucleation layer grown by migration-enhanced epitaxy is also adopted to suppress three-dimensional defects raised at the III-V/Si interface [70], which is followed by MT and HT GaAs to minimize the point defects during epitaxy [54]; Thermal cycle annealing (TCA) is another effective tool as it provides extra thermal stress to enhance the motion of TDs and promote higher probability for TD interaction [71].

In addition to the previously mentioned strategies, inserting SLSs serving as dislocation filter layers (DFLs) is another effective method that can sufficiently reduce the TDD. SLSs consist of periodic lattice-mismatched thin layers without strain relaxation. The unreleased strain tends to bend TDs at the SLSs interface and forces them to move laterally toward the edge of a crystal (parallel to the interface), enhancing the probability of intersection and termination of TDs, as shown in **Figure 13**. Besides, MD segments form at the SLSs interface as TD moves, relieving the misfit and reducing the net glide force of TDs to zero [43].

2.2.1 Optimization of SLSs and asymmetric step-graded filter structure

Optimizing SLSs to improve their filtering efficiency has been extensively explored. To design a proper SLS, the strain force must be first considered. A trade-off appears as a higher strain force strengthens the filtering ability, while an over-strain

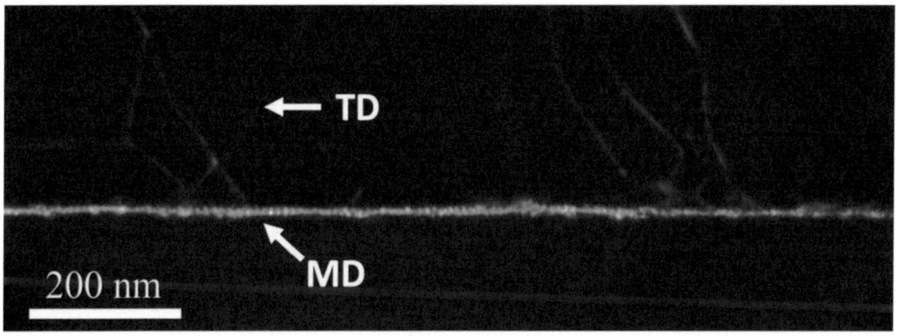

Figure 12.
Cross-sectional bright-field TEM image showing MDs and TDs.

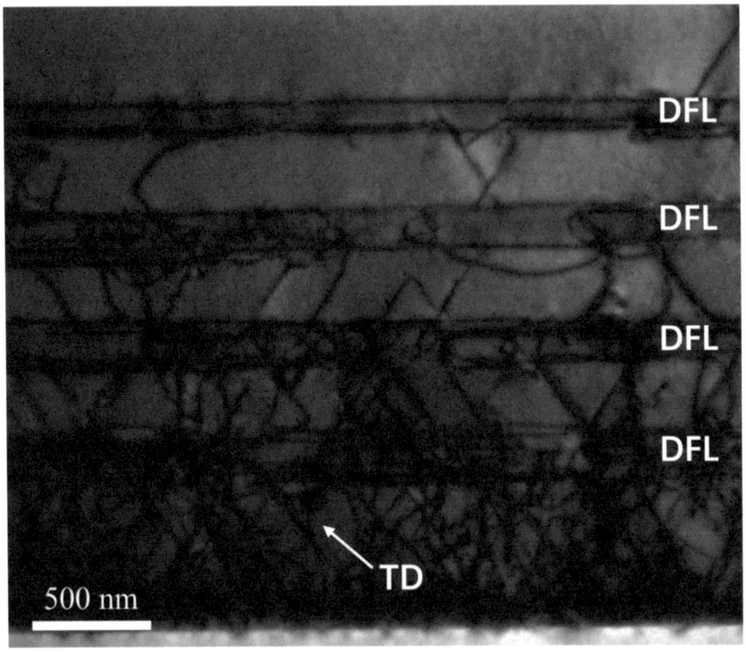

Figure 13.
Cross-sectional TEM image of GaAs grown on on-axis Si (001), showing TD propagation and annihilation within four sets of $In_{0.18}Ga_{0.82}As/GaAs$ DFLs.

force leads to new defects. Thus, for the most commonly used InGaAs/GaAs system, the indium (In) composition and the thickness of GaAs space layers for $In_xGa_{1-x}As$/GaAs DFLs must be carefully designed. Tang et al. demonstrated that 18% of indium (among 16, 18, and 20%), along with a 300 nm GaAs space layer, delivered the best filter efficiency in the design of $In_xGa_{1-x}As$/GaAs DFLs [72]. Besides, *in situ* thermal annealing was applied for each set of DFLs when the growth was being pulsed in the reactor. This approach further reduced the TDD, as the motion of TDs was enhanced, and thus, a higher possibility for TD self-annihilation was achieved.

Shang et al. provided a comprehensive study of optimizing DFLs on GaAs/GaP/Si (001) templates, which formed the basis for the high-temperature InAs/GaAs QD laser with an extrapolated lifetime of over 22 years [51, 71]. In this study, $In_{0.15}Ga_{0.85}As$ (10 nm)/GaAs (x nm) $\times$ 20 SLSs as DFLs were grown and compared. X represents values 10, 7, 5, 2, 0. A decreasing trend of filter efficiency was observed when x became lower, attributed to a higher degree of relaxation for $In_{0.15}Ga_{0.85}As$. Based on this observation, 200 nm $In_xGa_{1-x}As$ were further analyzed with various indium (In) compositions (10, 15, 17.5, 20, and 25%). It was concluded that 15 and 17.5% In composition delivered the best filtering efficiency, and a clear blocking effect of TDs was observed for an In composition higher than 20%. Furthermore, Shang et al. developed novel filter layers with ASG filter structure, which helps to reduce blocking effects, as shown in **Figure 14(a)**. Ten minutes of annealing at 530°C was applied for each InGaAs layer to promote tensile relaxation. Finally, TDD of 2 $\times 10^6$ cm^{-2} was achieved based on this structure, as shown in **Figure 14(b)**. This is also proven by TEM results shown in **Figure 14(c)**. Almost all TDs are blocked by this ASG filter structure, leaving the top GaAs layer with ultra-low TDD.

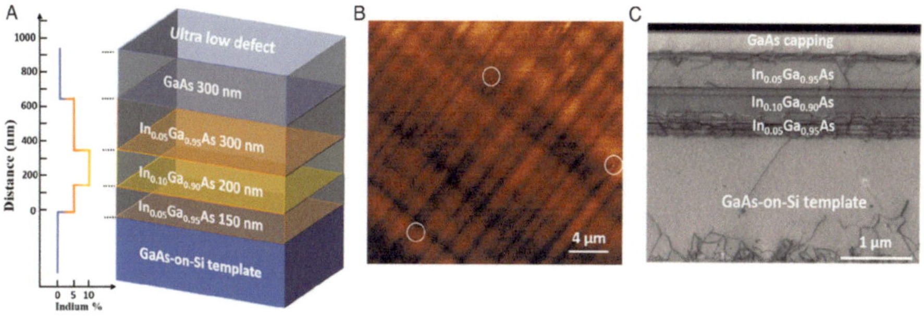

Figure 14.
(a) Schematic diagram of InGaAs ASG layers. (b) ECCI image showing TDD of 2 $\times 10^6$ cm^{-2}. (c) Cross-sectional TEM image of InGaAs ASG layers. Reprinted from [71] © 2020 the authors, under CC BY.

2.2.2 Trapping layer

During the postgrowth cooldown period, thermal stress-induced TD motion happens as a result of the large mismatch in CTE between III-V and Si. Once TDs encounter the In-contained DWELL structure, the mechanically hardened active region forces TDs to move laterally, leaving behind the MDs at the interface and severely degrading the optical properties of QDs. To solve this issue, Selvidge et al. inserted In-contained trapping layers (TL) to displace MDs above and below active region, which dramatically improved the optical prosperities of QDs [73]. The thickness of InGaAs TLs was kept below critical thickness (7 nm) without introducing extra MDs. As shown in **Figure 15**, inserting TLs does not contribute to TD reduction, but it displaces the MD formation below it rather than the active region to minimize the decremental impact brought by MDs. TLs were also applied in the laser structure to explore its efficiency, as illustrated in **Figure 16(a)**. 7 nm $In_{0.15}Ga_{0.15}As$ and $In_{0.15}Al_{0.15}As$ were placed 80 nm above and below the active region sandwiched by cladding layers to minimize the effect of the electrical barrier due to bandgap alignment. From **Figure 16(b)** and **(c)**, TLs were effective for displacing MDs along it rather than on DWELL. This observation is consistent with **Figure 16(c–f)**, where MDs lie on the DWELL when no TLs are inserted. Most MDs lie on the TLs, and further glide of TD segments does not introduce extra MDs in the QD lasers. It was

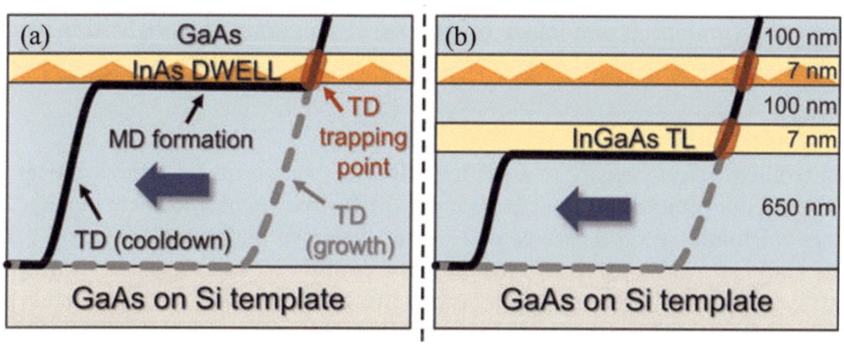

Figure 15.
Schematic diagrams showing MD formed (a) without and (b) with InGaAs TLs. Reprinted with permission from [73] ©2020 AIP publishing.

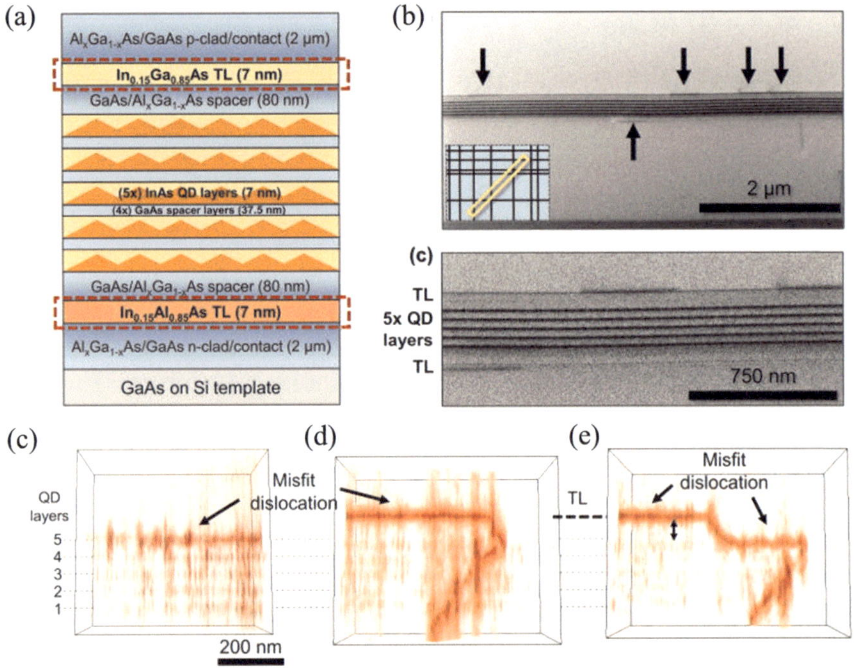

Figure 16.
(a) Schematic diagram of proposed laser structure with TLs. (b) Cross-sectional bright-field TEM image showing MD segments appear on TLs, as indicated by black arrows. (c) Zoom-in image of (b). Cross-sectional tomographic reconstruction showing (c) MDs lie on the 5th QDs in laser structure without TLs. (d) MDs lie on the TLs. (e) Part MDs lie on the TLs, and part lie on the 5th QD. Reprinted with permission from [73] ©2020 AIP publishing.

also demonstrated that the laser with TLs exhibited half of the threshold current (even lower than state-of-art lasers on Si when higher TDD is presented in this case), a 60% increase in slope efficiency, and ~ 3.4 times improvement in peak single facet output power, revealing the effectiveness of TLs. Such performance is comparable to Si-based QD lasers with one magnitude lower TDD. Compared with the structure consisting of thick DFLs, the insertion of thin TLs is more effective in improving laser performance without introducing a thick epilayer, which is beneficial for the yield and massive production of Si-based PICs in the long term.

Recent developments of reducing TDD in GaAs monolithically grown on Si (001) substrates by combining previously mentioned strategies are highlighted in **Table 1**.

2.3 Cracks

Controlling defect density to a sufficiently low value requires a thick buffer layer with several micrometers, which introduces the formation of micro-cracks. Because of the large mismatch in CTE between III-V materials and Si, for example, 5.73 $\times 10^{-6 o}C^{-1}$ for GaAs and 2.6 $\times$ $10^{-6 o}C^{-1}$ for Si, the accumulated thermal stress during the growth is relieved by forming micro-cracks and wafer warping when the epilayer cools down from high growth temperature to room temperature [79]. Theoretically, cracks form along the [110] and [1–10] directions when the elastic energy exceeds a critical value to generate two new surfaces, as shown in **Figure 17**. In addition, it is also proved that crack formation originates from other preexisting defects [80] and

Year	Substrate	Dislocation filter layer structure	Total buffer thickness (μm)	TDD (cm^{-2})	Ref
2021	V-grooved Si (001)	Two sets of five periods of $In_{0.15}Ga_{0.85}As/GaAs$ SLSs and two sets of five periods of $In_{0.15}Al_{0.85}As/GaAs$ SLSs	2.1	1.6×10^7	[74]
2020	GaP/Si (001)	$In_{0.05}Ga_{0.95}As/ In_{0.1}Ga_{0.9}As/In_{0.05}Ga_{0.95}As$ ASG	2.55	1.5×10^6	[71]
2020	Si (001)	Two sets of five periods of $In_{0.15}Ga_{0.85}As/GaAs$ SLSs	2.6	3×10^7	[75]
2020	GaP/Si (001)	Single $In_{0.1}Ga_{0.9}As$ layer	2.3	7.3×10^6	[76]
2019	GaAs/Si (001)	Four sets of five periods of $In_{0.18}Ga_{0.82}As/GaAs$ SLSs	2.3	5×10^7	[77]
2019	Si (001)	Three sets of five periods of $In_{0.15}Ga_{0.85}As/GaAs$ SLSs	2	Below 10^8	[55]
2017	GaP/Si (001)	Ten periods of $In_{0.1}Ga_{0.9}As/GaAs$ SLSs	3.1	8.4×10^6	[78]

Table 1.
Summary of recent optimization of DFL layers for GaAs monolithically grown on Si (001) substrates.

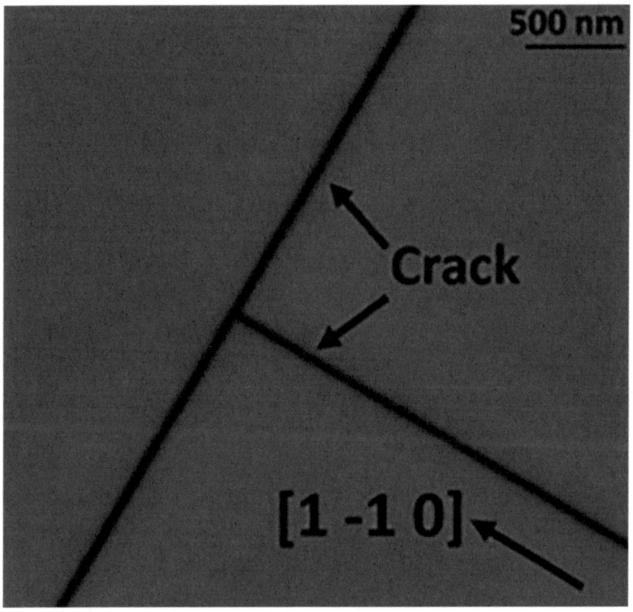

Figure 17.
SEM image of cracks in orthogonal directions. Reprinted from [79] © 2022 the authors, under CC BY.

layer thickness. For example, Yang et al. reported that the crack density increases sharply about three times when the thickness of the GaAs layer on Si increases from 5 to 6.7 μm [81]. Similar to other defects, micro-cracks are detrimental to the device's performance as they serve as scattering centers for light propagation and electrical leakage paths [81]. Additionally, the high density of micro-cracks will significantly reduce the total yield of devices [82]. Hence, controlling micro-crack is crucial for the mass production of Si-based PICs in the future.

A prolonged cooling down period with a slower cooling rate is suggested after growth to prevent the micro-cracks [79, 83]. Furthermore, SAG of III-V materials helps to prevent micro-cracks formation by alleviating thermal stress. However, dense defects, including TDs and stacking faults, will be generated near the pattern edge, degrading crystal quality. Moreover, in large patterned areas, micro-cracks remain on the sample surface [84]. Even though diverse techniques have been demonstrated, keeping the device thickness below the cracking threshold is the most economical and effective way [81].

As cracks are formed when the elastic energy exceeds a certain limitation, the thickness of the epilayers is the most prominent and essential reason for the crack formation. Yang et al. proposed the relationship between the critical cracking thickness and a dimensionless driving force number Z [85], which calculates the energy released per unit area for the crack:

$$G = \frac{Z\sigma^2 t}{\overline{E_f}} \tag{3}$$

where σ is the stress in the thin film, t is the thin film thickness and $\overline{E_f}$ denotes the biaxial modulus. In a typical system where the substrate and the epilayer have similar elastic moduli, the Z should be within 2 to 4. The stress in the thin film can be calculated as:

$$\sigma = \overline{E_f}\left(\alpha_f - \alpha_s\right)\Delta T \tag{4}$$

where α_f and α_s are the CTE of the thin film and the substrate, respectively. While the ΔT is the temperature difference between the growth temperature and the room temperature. The critical thickness for crack formation can be derived, provided that the fracture resistance Γ is twice the energy release rate G [85, 86]:

$$t_c = \frac{\Gamma \overline{E_f}}{Z\sigma^2} \tag{5}$$

Based on the mathematical model given above, the cracking threshold of GaAs is estimated as 3.9 μm when it is cooled down from the growth temperature of around 600°C to room temperature. However, for most growth of high-performance III-V compound semiconductor devices on Si, the structure was usually above 4 μm due to the utilization of a thick buffer layer to minimize the TD generated at the interface.

Furthermore, Shang et al. further improved the previous model and shed light on the relationship among dislocation density, film thickness, cooling rate, and crack formation [79]. It has been suggested that lower TDD induces higher equi-biaxial stress in the film during the cooling down period. The critical thickness is inversely proportional to the cool rate and TDD, as shown in **Figure 18**. It is suggested that with a low TDD of 1.0×10^6 cm^{-2} and a low cool rate of 1°C min^{-1}, the critical thickness of cracking is approximately 6 μm. Therefore, a thin epilayer with a low cooling rate is of pinnacle importance to prevent micro-cracks.

Recently, Yang et al. used an optimized 300 nm Ge buffer layer to replace part of the thick GaAs buffer layer in the laser structure while keeping the TDD unchanged [82, 87]. As a result, the total thickness of the laser structure can be reduced to approach the cracking threshold without bringing any negative effects. A comparison between TD propagation for GaAs deposited directly on a Si substrate and a Ge/Si

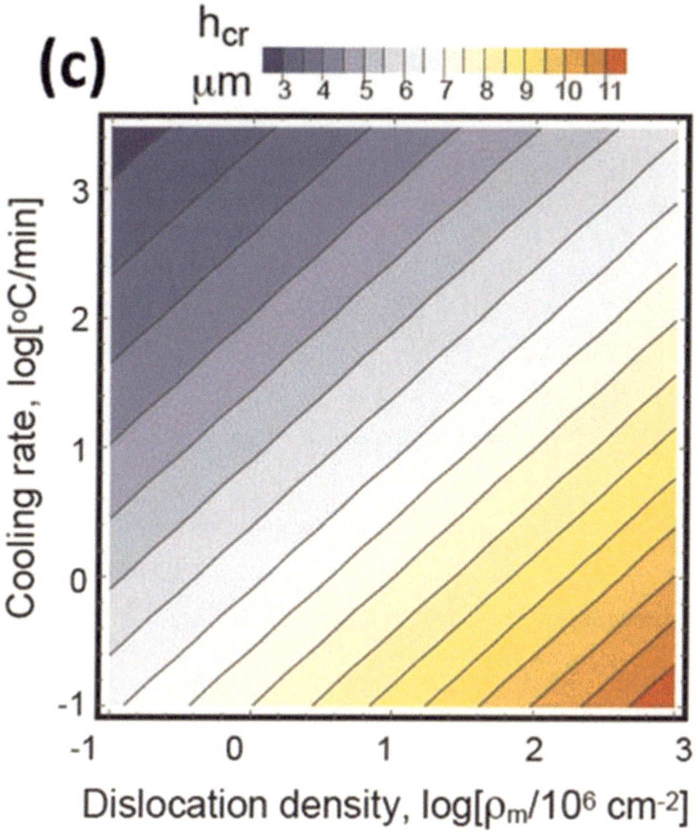

Figure 18.
Relationship among critical thickness, cooling rate, and TDD. Reprinted from [79] © 2022 the authors, under CC BY.

virtual substrate (VS) is demonstrated in **Figure 19**. As shown in **Figure 19(a)**, a high density of defects is generated at the GaAs/Si interface, and almost $\sim 10^9$ cm^{-2} TDD is observed underneath the first DFL [54]. In contrast, a much lower TDD of $\sim 6 \times 10^8$ cm^{-2} is obtained in the 300 nm Ge buffer layer attributed to the adoption of HT TCA between 600 and 900°C, as indicated in **Figure 19(b)**. This TDD is comparable to the 1.4 μm GaAs monolithically grown on Si with one set of DFL. In addition, MDs are barely introduced in the subsequent GaAs growth since the lattice constants of Ge and GaAs are almost identical. As a result, TDD reaches 4×10^6 cm^{-2} after applying four sets of DFLs. Based on this result, a high-quality InAs/GaAs QD laser was developed with high operation temperature, revealing the feasibility of using Ge/Si platform for reducing micro-cracks in the future.

2.4 Summary

Recent progress in controlling crystal defects during the heteroepitaxy of III-V materials on Si substrates has been reviewed in this section. Several newly developed techniques were applied for Si-based PICs, which will be discussed in the following contents.

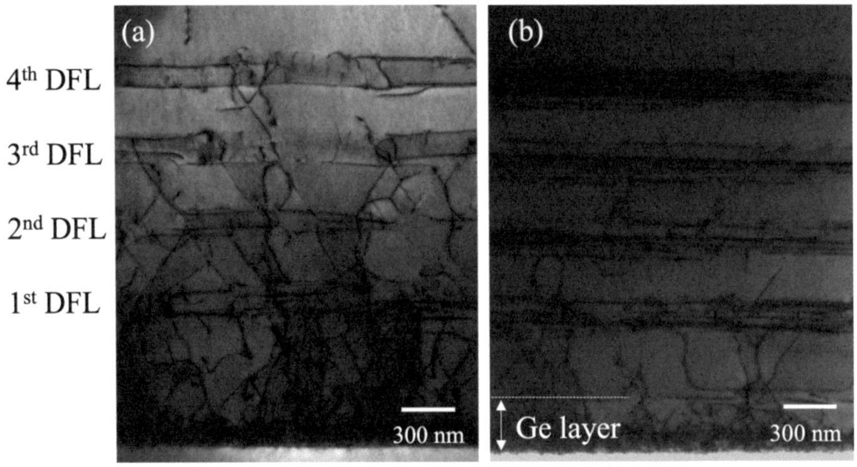

Figure 19.
Cross-sectional TEM images of GaAs buffer layer grown on (a) a Si substrate and (b) a Ge/Si VS. reprinted from [82] © 2021 the authors, under CC BY.

3. Photonic integrated circuits

The idea of using Si-based PICs in which all major photonic functions are monolithically integrated on a single Si or Si-on-insulator (SOI) substrate has emerged to promote rapid advances in quantum photonics, quantum computing, LiDAR, and artificial intelligence-powered nanophotonics [4, 88]. It contributes to the better life quality of consumers with low cost due to the low material cost and large wafer size of Si [4, 89, 90]. Over the past decades, an unprecedented boom of key components of Si photonics, including Si-based modulators [91], photodetectors [92], and waveguides [93], has been witnessed. Until now, an efficient, electrically pumped Si-based laser remains a missing piece and becomes the roadblock to the commercialization of Si-based PICs.

To circumvent the inherent limitations of Si, integrating direct-bandgap III-V materials onto Si has been regarded as an attractive approach for the Si-based on-chip light source in PICs. Such integration leverages the benefits of superior optical properties of III-V materials, along with large wafer sizes and the low-cost and mature processing technology of Si. Direct epitaxy of QD-based laser on Si substrates has achieved remarkable progress [51, 54, 94, 95]. Various novel laser structures were reported with superior performance, such as distributed feedback lasers [96], comb lasers [97], photonic crystal lasers [98], topological lasers [99], etc. All key optical components integrated on a single SOI substrate are highly desired as they offer high integration density and great compatibility with the current Si microelectronics platform. Until now, the integration of III-V gain regions on SOI substrates mainly relies on wafer bonding, in which light is evanescently coupled to underlying Si waveguides [9, 20]. However, from the commercialization perspective, direct epitaxy is economically favored in terms of cost, yield, and scalability. Considering the integration of on-chip laser sources, the thick, defective buffer layer adopted for direct epitaxy of III-V materials on Si hinders the evanescent coupling of light from gain regions to underlying Si waveguides. In this case, SAG growth of laser structure on a trenched substrate and butt-coupled to the embedded, prepatterned waveguide is promising for fulfilling the last missing piece of Si-based PICs. In addition, SAG helps to alleviate the thermal stress of films, potentially preventing the formation of cracks [79]. On the other hand, such a method is nontrivial as it demands restricted design to minimize the alignment deviation between the central axis of the embedded Si waveguide and the InAs/GaAs QD active region and careful handle of polycrystal after the overgrowth of III-V on oxide.

Shang et al. recently reported the first electrically pumped continuous-wave (c.w.) InAs/GaAs QD lasers grown on a patterned 300 mm substrate [100]. In this study, the 200 nm GaP/Si template by NAsP $_{III/V}$ GmbH was adopted in trenches to prevent APBs [47, 66]. This was followed by a 1.6 μm GaAs buffer layer and InGaAs asymmetric graded dislocation filter layers to reduce TDD to around 1.5×10^7 cm^{-2}. The active region consists of five stacks of InAs/GaAs DWELL structure separated by 37.5 nm GaAs space layers. The QD nucleation temperature was precisely determined using indium as an ex situ "temperature gauge" in this template to ensure high-quality QDs. As a result, room temperature with a wavelength of 1300 nm and a full width at half maximum (FWHM) of 32 meV was achieved for QDs. The as-grown 300 mm wafer from IQE with an identical growth method is shown in **Figure 20(a)**. The milky wafer surface caused by the deposited polycrystalline III-V on the oxide brought challenging tasks for device fabrication. A wet etch process for nonselective polycrystal removal facilitates the following fabrication process, as shown in **Figure 20(b)**. The top-down view of the as-cleaved laser with probe metal and the cross-sectional SEM image of the fabricated laser with a ridge width of 3.5 μm in a 20 μm trench are demonstrated in **Figure 20(c)** and **(d)**, respectively.

Finally, an electrically pumped InAs/GaAs laser was demonstrated with c.w. lasing up to 60°C, a maximum double-side power of 126.6 mW, and a threshold current of 47.5 mA. However, this work only presents a demo of an in-trench laser without demonstrating butt coupling between the laser and the embedded waveguide.

In a parallel effort, Wei et al. took a step further to test the butt coupling efficiency between their embedded InAs/GaAs QD lasers and Si waveguides. **Figure 21(a)** shows a schematic diagram of butt coupling between a trenched laser and a patterned Si waveguide. The fabricated devices are displayed in **Figure 21(b)** and **(c)**. Prior to growth, laser trenches and Si waveguides are prepatterned in an eight-inch SOI wafer,

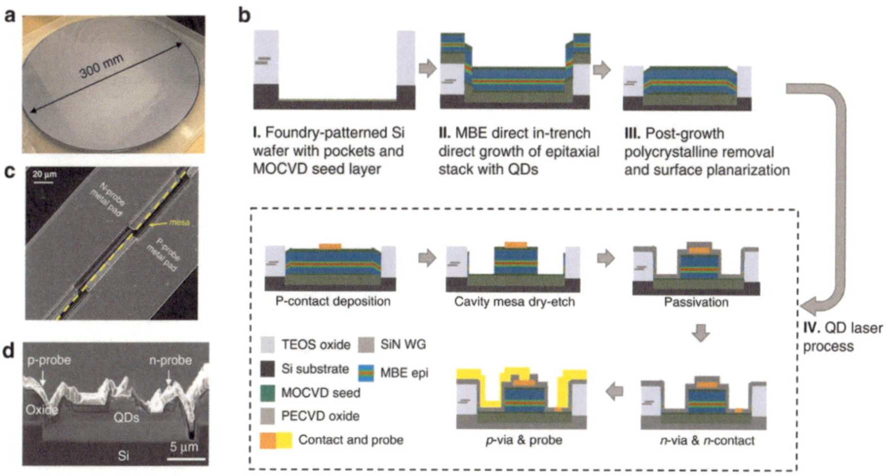

Figure 20.
(a) As-grown 300 mm wafer surface. (b) the fabrication process of the SAG laser. (c) As-cleaved laser with probe metal. (d) Cross-sectional SEM image of the fabricated laser in a 20 μm trench. ©2022 the authors under CC BY.

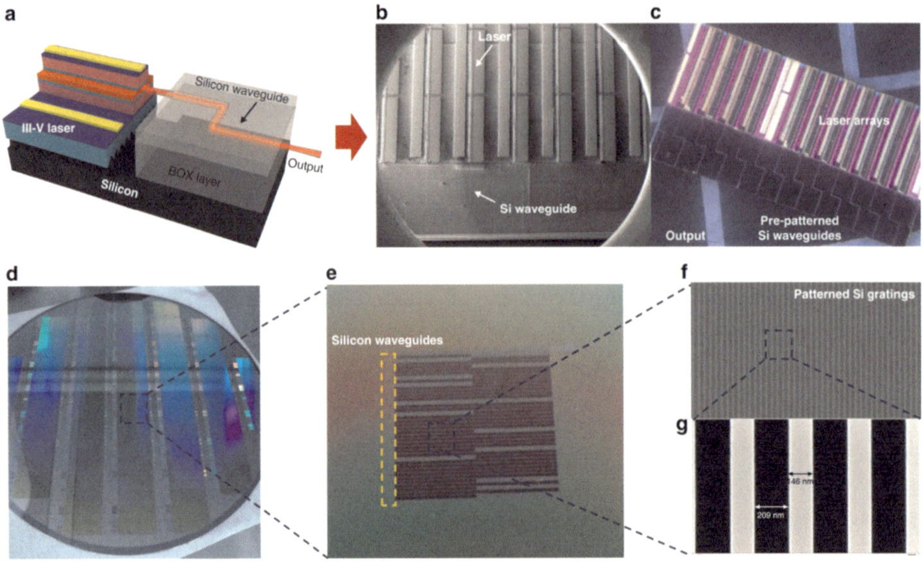

Figure 21.
(a) Schematic image of butt coupling between laser structure and Si waveguide. (b), (c) SEM and microscope images of trenched InAs/GaAs laser structure with prepatterned Si waveguide. (d)eight-inch wafer with predefined trenches and waveguides. (e) Microscope images of trenches for laser epitaxy and embedded Si waveguide. (f) and (d) patterned Si grating with 146 nm slab width and 209 nm gap for Si homoepitaxy and form Si {111} surfaces. ©2023 the authors under CC BY.

and the periodic Si gratings are patterned inside the trench with 146 nm slab width and a 209 nm gap, as demonstrated in **Figure 21(d)** and **(e)**.

Instead of using a commercially available 200 nm GaP/Si template, Wei et al. adopted homoepitaxy of Si on the grating-patterned SOI trenches, which forms Si {111} facets to prevent the formation of APBs [46, 101]. The trenched laser is demonstrated in **Figure 22(a)**. A combination of a thin AlAs nucleation layer, a 2.1 μm GaAs

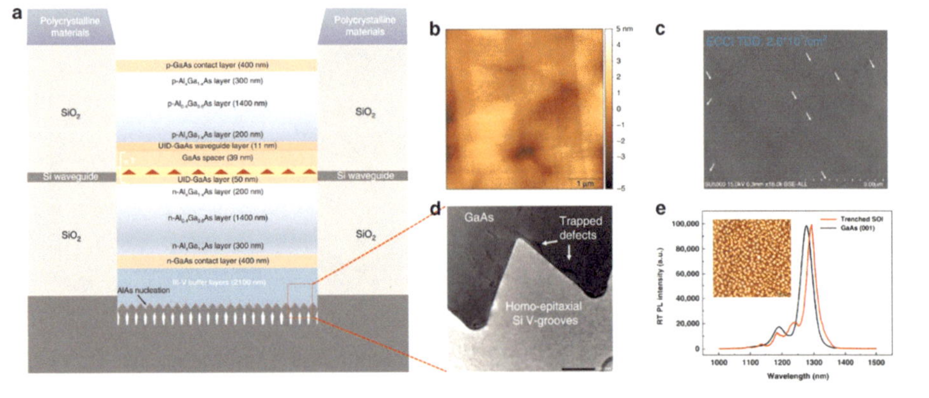

Figure 22.
(a) Schematic image of the trenched laser structure. (b) 5 × 5 μm² AFM image of 2.1 μm GaAs buffer layer. (c) TDD of 2.6 × 10⁷ cm⁻² is obtained for the GaAs buffer layer. (d) Cross-sectional TEM image of GaAs/Si (111) interface. (e) Comparison of PL measurement of trenched QD laser and blanket GaAs (001) laser with identical growth structure. Inset: Surface morphology of grown InAs/GaAs QDs of trenched laser. ©2023 the authors under CC BY.

buffer layer consisting of InGaAs/GaAs DFLs, and GaAs/AlAs SLSs was adopted to reduce TDD to 2.6×10^7 cm^{-2} while maintaining low surface roughness of 0.8 nm in a 5×5 μm^2 AFM scan, as shown in **Figure 22(b)–(d)**. The active region consists of seven stacks of InAs/GaAs DWELL structure separated by 39 nm GaAs space layers, sandwiched by 400 nm GaAs contact layers and $Al_{0.4}Ga_{0.6}As$ cladding layers. A step-graded AlGaAs layers were also adopted to enhance the current injection efficiency. A comparison of PL measurements between a laser with an identical structure grown on a trench and a GaAs (001) is given in **Figure 22(e)**. A similar PL intensity with a narrow FWHM of 33 nm is observed for trenched laser, attributed to high density and high uniform QDs, as shown in the inset of **Figure 22(e)**.

In this study, H_3PO_4:H_2O_2:H_2O (1:2:20) wet etching was applied to remove unwanted polycrystalline III-V materials before fabrication. The trenched laser was processed with one-side cleaved and coated with a high-reflection coating. While the other side implements wet etch followed by two-step focused ion beam milling to produce high-quality facets. A High-performance trenched QD laser was fabricated with c.w. lasing up to 85°C, low threshold current of 50 mA and maximum output power of 37 mW at an injection current of 250 mA. Butt coupling efficiency between QD active region and Si waveguide was determined. A maximum power of 6.8 mW was measured at the end tip of the Si waveguide, indicating ~ -6.7db coupling efficiency. A further improvement in coupling efficiency can be achieved by using an advanced silicon spot-size converter with precise control of the gap between the facet and the waveguide. This laser offers a prospective technique for realizing an on-chip light source for Si-based PICs.

3.1 Summary

Demonstrating high-performance trenched QD lasers shapes the faith in realizing the monolithic integration of III-V lasers for Si-based PICs as on-chip sources. It paves the way toward large-scale, high-density, low-cost PICs for the forthcoming bloom of quantum and sensing technologies.

4. Conclusions

Heteroepitaxial growth of III-V materials onto Si substrates offers an appealing approach for achieving practical Si-based on-chip light sources. Because of the large lattice mismatch between III-V materials and Si, the formation of crystal defects, including TDs, APBs, and micro-cracks, is inevitable during the epitaxy. Over the past decades, great advances in growth techniques have been made to control these defects to a reasonably low value, and state-of-the-art techniques are reviewed in this chapter. Hence, the performance of InAs QD laser grown on Si substrates progresses rapidly in terms of threshold current, maximum working temperature, and reliability. A step further is urgent to migrate these techniques into Si-based PICs, which are primed to support the growing market of automotive, sensing techniques, and quantum technologies. In order to integrate the InAs/GaAs QD light source on Si-based PICs, SAG growth of laser structure on a trenched substrate and butt-coupled to the embedded, prepatterned waveguide is regarded as a promising candidate for realizing on-chip light sources. Though only a few reports have demonstrated a demo for coupling light from trenched InAs QDs active region into the waveguide, the initial results are promising, which shapes the faith of achieving monolithic integration of III-V lasers as an on-chip light source. The realization of Si-based PICs will undoubtedly unleash the great potential of emerging technologies in the near future.

Acknowledgements

This work was supported by the UK Engineering and Physical Sciences Research Council (EP/P006973/1, EP/T028475/1, EP/X015300/1).

Author details

Junjie Yang*, Huiwen Deng, Jae-Seong Park, Siming Chen, Mingchu Tang and Huiyun Liu
Department of Electronic and Electrical Engineering, University College London, London, United Kingdom

*Address all correspondence to: zceejya@ucl.ac.uk

References

[1] Soref R. The past, present, and future of silicon photonics. IEEE Journal of Selected Topics in Quantum Electronics. 2006;**12**(6):1678-1687

[2] Jalali B, Fathpour S. Silicon photonics. Journal of Lightwave Technology. 2006; **24**(12):4600-4615

[3] Tang M, Park J-S, Wang Z, Chen S, Jurczak P, Seeds A, et al. Integration of III-V lasers on Si for Si photonics. Progress in Quantum Electronics. 2019; **66**:1-18

[4] Zhou Z, Ou X, Fang Y, Emad A, Xu R, Yating W, et al. Prospects and applications of on-chip lasers. eLight. 2023;**3**(1):1

[5] Wang Z, Junesand C, Metaferia W, Hu C, Wosinski L, Lourdudoss S. III–vs on Si for photonic applications—A monolithic approach. Materials Science and Engineering: B. 2012;**177**(17): 1551-1557

[6] Kunert B, Guo W, Mols Y, Tian B, Wang Z, Shi Y, et al. III/V nano ridge structures for optical applications on patterned 300 mm silicon substrate. Applied Physics Letters. 2016;**109**(9): 091101

[7] Norman JC, Jung D, Wan Y, Bowers JE. Perspective: The future of quantum dot photonic integrated circuits. APL Photonics. 2018;**3**(3): 030901

[8] Norman JC, Jung D, Zhang Z, Wan Y, Liu S, Shang C, et al. A review of high-performance quantum dot lasers on silicon. IEEE Journal of Quantum Electronics. 2019;**55**(2):1-11

[9] Roelkens G, Liu L, Liang D, Jones R, Fang A, Koch B, et al. III-V/silicon photonics for on-chip and intra-chip optical interconnects. Laser & Photonics Reviews. 2010;**4**(6):751-779

[10] Liu H, Wang T, Jiang Q, Hogg R, Tutu F, Pozzi F, et al. Long-wavelength InAs/GaAs quantum-dot laser diode monolithically grown on Ge substrate. Nature Photonics. 2011;**5**(7):416

[11] Liang D, Bowers JE. Recent progress in lasers on silicon. Nature Photonics. 2010;**4**(8):511-517

[12] Mi Z, Yang J, Bhattacharya P, Qin G, Ma Z. High-performance quantum dot lasers and integrated optoelectronics on Si. Proceedings of the IEEE. 2009;**97**(7): 1239-1249

[13] Lee A, Jiang Q, Tang M, Seeds A, Liu H. Continuous-wave InAs/GaAs quantum-dot laser diodes monolithically grown on Si substrate with low threshold current densities. Optics Express. 2012; **20**(20):22181-22187

[14] Liu AY, Zhang C, Snyder A, Lubyshev D, Fastenau JM, Liu AWK, et al. MBE growth of P-doped 1.3μm InAs quantum dot lasers on silicon. Journal of Vacuum Science & Technology B, Nanotechnology and Microelectronics: Materials, Processing, Measurement, and Phenomena. 2014;**32**(2):02C108

[15] Deng H, Yang J, Jia H, Tang M, Maglio B, Jarvis L, et al. Si-Based 1.3 μm InAs/GaAs QD Lasers. 2022 IEEE Photonics Conference (IPC), Vancouver. IEEE; 2022

[16] Liao M, Chen S, Park J-S, Seeds A, Liu H. III–V quantum-dot lasers monolithically grown on silicon. Semiconductor Science and Technology. 2018;**33**(12):123002

[17] Wang T, Liu H, Lee A, Pozzi F, Seeds A. 1.3-μm InAs/GaAs quantum-dot lasers monolithically grown on Si substrates. Optics Express. 2011;**19**(12): 11381-11386

[18] Tang M, Chen S, Wu J, Jiang Q, Dorogan VG, Benamara M, et al. 1.3-μm InAs/GaAs quantum-dot lasers monolithically grown on Si substrates using InAlAs/GaAs dislocation filter layers. Optics Express. 2014;**22**(10):11528-11535

[19] Park H, Fang AW, Kodama S, Bowers JE. Hybrid silicon evanescent laser fabricated with a silicon waveguide and III-V offset quantum wells. Optics Express. 2005;**13**(23):9460-9464

[20] Fang AW, Park H, Cohen O, Jones R, Paniccia MJ, Bowers JE. Electrically pumped hybrid AlGaInAs-silicon evanescent laser. Optics Express. 2006; **14**(20):9203-9210

[21] Chang H-H, Fang AW, Sysak MN, Park H, Jones R, Cohen O, et al. 1310 nm silicon evanescent laser. Optics Express. 2007;**15**(18):11466-11471

[22] Fang AW, Koch BR, Jones R, Lively E, Liang D, Kuo YH, et al. A distributed Bragg reflector silicon evanescent laser. IEEE Photonics Technology Letters. 2008;**20**(20): 1667-1669

[23] Liang D, Roelkens G, Baets R, Bowers J. Hybrid integrated platforms for silicon photonics. Materials. 2010; **3**(3):1782

[24] Mickael M, Thierry B, Yann B, Huiwen D, Keshuang L, Mingchu T, et al. Ch. 8. GaAs compounds heteroepitaxy on silicon for Opto and nano electronic applications. In: Mohammed Muzibur R, Abdullah Mohammed A, Anish K, Inamuddin TT, editors. Post-Transition Metals. Rijeka: IntechOpen; 2020

[25] Kunert B, Mols Y, Baryshniskova M, Waldron N, Schulze A, Langer R. How to control defect formation in monolithic III/V hetero-epitaxy on (100) Si? A critical review on current approaches. Semiconductor Science and Technology. 2018;**33**(9):093002

[26] Park J-S, Tang M, Chen S, Liu H. Heteroepitaxial growth of III-V semiconductors on silicon. Crystals. 2020;**10**(12):1163

[27] Alerhand O, Berker AN, Joannopoulos J, Vanderbilt D, Hamers R, Demuth J. Finite-temperature phase diagram of vicinal Si (100) surfaces. Physical Review Letters. 1990;**64**(20): 2406

[28] Pehlke E, Tersoff J. Phase diagram of vicinal Si(001) surfaces. Physical Review Letters. 1991;**67**(10):1290-1293

[29] Poon TW, Yip S, Ho PS, Abraham FF. Equilibrium structures of Si (100) stepped surfaces. Physical Review Letters. 1990;**65**(17):2161

[30] De Miguel J, Aumann C, Kariotis R, Lagally M. Evolution of vicinal Si (001) from double-to single-atomic-height steps with temperature. Physical Review Letters. 1991;**67**(20): 2830

[31] Chadi DJ. Stabilities of single-layer and bilayer steps on Si(001) surfaces. Physical Review Letters. 1987;**59**(15): 1691-1694

[32] Yang J. Monolithic Growth of InAs/GaAs Quantum Dot Lasers on Silicon Substrates by Molecular Beam Epitaxy [thesis]. England: University College London; 2021

[33] Kroemer H. Polar-on-nonpolar epitaxy. Journal of Crystal Growth. 1987; **81**(1–4):193-204

[34] Martin M, Caliste D, Cipro R, Alcotte R, Moeyaert J, David S, et al. Toward the III–V/Si co-integration by controlling the biatomic steps on hydrogenated Si(001). Applied Physics Letters. 2016;**109**(25):253103

[35] Alcotte R, Martin M, Moeyaert J, Cipro R, David S, Bassani F, et al. Epitaxial growth of antiphase boundary free GaAs layer on 300 mm Si(001) substrate by metalorganic chemical vapour deposition with high mobility. APL Materials. 2016;**4**(4):046101

[36] Rubel O, Baranovskii SD. Formation energies of antiphase boundaries in GaAs and GaP: An ab initio study. International Journal of Molecular Sciences. 2009;**10**(12):5104-5114

[37] Beyer A, Volz K. Advanced electron microscopy for III/V on silicon integration. Advanced Materials Interfaces. 2019;**6**(12):1801951

[38] Chen S, Liao M, Tang M, Wu J, Martin M, Baron T, et al. Electrically pumped continuous-wave 1.3 microm InAs/GaAs quantum dot lasers monolithically grown on on-axis Si (001) substrates. Optics Express. 2017; **25**(5):4632-4639

[39] Bai J, Park J-S, Cheng Z, Curtin M, Adekore B, Carroll M, et al. Study of the defect elimination mechanisms in aspect ratio trapping Ge growth. Applied Physics Letters. 2007;**90**(10): 101902

[40] Yan Z, Han Y, Lin L, Xue Y, Ma C, Ng WK, et al. A monolithic InP/SOI platform for integrated photonics. Light: Science & Applications. 2021;**10**(1):200

[41] Krost A, Schnabel R, Heinrichsdorff F, Rossow U, Bimberg D, Cerva H. Defect reduction in GaAs and InP grown on planar Si (111) and on patterned Si (001) substrates. Journal of Crystal Growth. 1994;**145**(1–4):314-320

[42] Fang S, Adomi K, Iyer S, Morkoc H, Zabel H, Choi C, et al. Gallium arsenide and other compound semiconductors on silicon. Journal of Applied Physics. 1990; **68**(7):R31-R58

[43] Beanland R, Dunstan D, Goodhew P. Plastic relaxation and relaxed buffer layers for semiconductor epitaxy. Advances in Physics. 1996;**45**(2):87-146

[44] Li Q, Lau KM. Epitaxial growth of highly mismatched III-V materials on (001) silicon for electronics and optoelectronics. Progress in Crystal Growth and Characterization of Materials. 2017;**63**(4):105-120

[45] Li Q, Ng KW, Lau KM. Growing antiphase-domain-free GaAs thin films out of highly ordered planar nanowire arrays on exact (001) silicon. Applied Physics Letters. 2015;**106**(7):072105

[46] Wei W-Q, Wang J-H, Zhang B, Zhang J-Y, Wang H-L, Feng Q, et al. InAs QDs on (111)-faceted Si (001) hollow substrates with strong emission at 1300 nm and 1550 nm. Applied Physics Letters. 2018;**113**(5):053107

[47] Kunert B, Nemeth I, Reinhard S, Volz K, Stolz W. Si (001) surface preparation for the antiphase domain free heteroepitaxial growth of GaP on Si substrate. Thin Solid Films. 2008;**517**(1): 140-143

[48] Beyer A, Németh I, Liebich S, Ohlmann J, Stolz W, Volz K. Influence of crystal polarity on crystal defects in GaP grown on exact Si (001). Journal of Applied Physics. 2011;**109**(8):083529

[49] Kunert B, Volz K. Monolithic III/V integration on (001) Si substrate. In: Metalorganic Vapor Phase Epitaxy

(MOVPE) Growth, Materials Properties, and Applications. Chichester: John Wiley & Sons; 2019. pp. 241-291

[50] Jung D, Norman J, Kennedy MJ, Shang C, Shin B, Wan Y, et al. High efficiency low threshold current 1.3μm InAs quantum dot lasers on on-axis (001) GaP/Si. Applied Physics Letters. 2017;**111**(12):122107

[51] Shang C, Hughes E, Wan Y, Dumont M, Koscica R, Selvidge J, et al. High-temperature reliable quantum-dot lasers on Si with misfit and threading dislocation filters. Optica. 2021;**8**(5): 749-754

[52] Lv Z, Wang S, Wang S, Chai H, Meng L, Yang X, et al. Ultra-high thermal stability InAs/GaAs quantum dot lasers grown on on-axis Si (001) with a record-high continuous-wave operating temperature of 150°C. Optics Express. 2023;**31**(15):24173-24182

[53] Jia H, Yang J, Tang M, Li W, Jurczak P, Yu X, et al. The epitaxial growth and unique morphology of InAs quantum dots embedded in a Ge matrix. Journal of Physics D: Applied Physics. 2022;**55**(49): 494002

[54] Chen S, Li W, Wu J, Jiang Q, Tang M, Shutts S, et al. Electrically pumped continuous-wave III–V quantum dot lasers on silicon. Nature Photonics. 2016;**10**(5):307-311

[55] Kwoen J, Jang B, Lee J, Kageyama T, Watanabe K, Arakawa Y. All MBE grown InAs/GaAs quantum dot lasers on on-axis Si (001). Optics Express. 2018; **26**(9):11568-11576

[56] Kwoen J, Jang B, Watanabe K, Arakawa Y. High-temperature continuous-wave operation of directly grown InAs/GaAs quantum dot lasers on on-axis Si (001). Optics Express. 2019; **27**(3):2681-2688

[57] Li K, Yang J, Lu Y, Tang M, Jurczak P, Liu Z, et al. Inversion boundary annihilation in GaAs monolithically grown on on-axis silicon (001). Advanced Optical Materials. 2020;**8**(22):2000970

[58] Yang J, Li K, Jia H, Deng H, Yu X, Jurczak P, et al. Low threading dislocation density and antiphase boundary free GaAs epitaxially grown on on-axis Si (001) substrates. Nanoscale. 2022;**14**(46):17247-17253

[59] Kwoen J, Lee J, Watanabe K, Arakawa Y. Elimination of anti-phase boundaries in a GaAs layer directly-grown on an on-axis Si(001) substrate by optimizing an AlGaAs nucleation layer. Japanese Journal of Applied Physics. 2019;**58**(SB):SBBE07

[60] Guo W, Bondi A, Cornet C, Létoublon A, Durand O, Rohel T, et al. Thermodynamic evolution of antiphase boundaries in GaP/Si epilayers evidenced by advanced X-ray scattering. Applied Surface Science. 2012;**258**(7): 2808-2815

[61] Gilbert A, Ramonda M, Cerutti L, Cornet C, Patriarche G, Tournié É, et al. Epitaxial growth of III-vs on on-axis Si: breaking the symmetry for antiphase domains control and burying. Advanced Optical Materials. 2023;**11**(15):2203050

[62] Tersoff J, Pehlke E. Sinuous step instability on the Si (001) surface. Physical Review Letters. 1992;**68**(6):816

[63] Tromp R, Reuter M. Wavy steps on Si (001). Physical Review Letters. 1992; **68**(6):820

[64] Sakamoto T, Kawai NJ, Nakagawa T, Ohta K, Kojima T. Intensity oscillations

of reflection high-energy electron diffraction during silicon molecular beam epitaxial growth. Applied Physics Letters. 1985;**47**(6):617-619

[65] Sakamoto T, Kawamura T, Hashiguchi G. Observation of alternating reconstructions of silicon (001) 2×1 and 1×2 using reflection high-energy electron diffraction during molecular beam epitaxy. Applied Physics Letters. 1986;**48**(23):1612-1614

[66] Volz K, Beyer A, Witte W, Ohlmann J, Németh I, Kunert B, et al. GaP-nucleation on exact Si (001) substrates for III/V device integration. Journal of Crystal Growth. 2011;**315**(1): 37-47

[67] Horikoshi Y, Kawashima M, Yamaguchi H. Migration-enhanced epitaxy of GaAs and AlGaAs. Japanese Journal of Applied Physics. 1988;**27**(2R): 169

[68] Ohta K, Kojima T, Nakagawa T. Anisotropic surface migration of Ga atoms on GaAs (001). Journal of Crystal Growth. 1989;**95**(1–4):71-74

[69] Groenert ME, Leitz CW, Pitera AJ, Yang V, Lee H, Ram RJ, et al. Monolithic integration of room-temperature cw GaAs/AlGaAs lasers on Si substrates via relaxed graded GeSi buffer layers. Journal of Applied Physics. 2003;**93**(1): 362-367

[70] Lee AD, Jiang Q, Tang M, Zhang Y, Seeds AJ, Liu H. InAs/GaAs quantum-dot lasers monolithically grown on Si, Ge, and Ge-on-Si substrates. IEEE Journal of Selected Topics in Quantum Electronics. 2013;**19**:4, 1901107

[71] Shang C, Selvidge J, Hughes E, Norman JC, Taylor AA, Gossard AC, et al. A pathway to thin GaAs virtual substrate on on-axis Si (001) with ultralow threading dislocation density. Physica Status Solidi (a). 2021;**218**(3): 2000402

[72] Tang M, Chen S, Wu J, Jiang Q, Kennedy K, Jurczak P, et al. Optimizations of defect filter layers for 1.3-μm InAs/GaAs quantum-dot lasers monolithically grown on Si substrates. IEEE Journal of Selected Topics in Quantum Electronics. 2016;**22**(6):50-56

[73] Selvidge J, Norman J, Hughes ET, Shang C, Jung D, Taylor AA, et al. Defect filtering for thermal expansion induced dislocations in III–V lasers on silicon. Applied Physics Letters. 2020;**117**(12): 122101

[74] Wei W, Huang J, Ji Z, Han D, Yang B, Chen J, et al. Reliable InAs quantum dot lasers grown on patterned Si (001) substrate with embedded hollow structures assisted thermal stress relaxation. Journal of Physics D: Applied Physics. 2022;**55**(40):405105

[75] Wan Y, Shang C, Norman J, Shi B, Li Q, Collins N, et al. Low threshold quantum dot lasers directly grown on unpatterned quasi-nominal (001) Si. IEEE Journal of Selected Topics in Quantum Electronics. 2020;**26**(2):1-9

[76] Wan Y, Norman JC, Tong Y, Kennedy M, He W, Selvidge J, et al. 1.3 μm quantum dot-distributed feedback lasers directly grown on (001) Si. Laser & Photonics Reviews. 2020;**14**(7): 2000037

[77] Liu Z, Hantschmann C, Tang M, Lu Y, Park J-s, Liao M, et al. Origin of defect tolerance in InAs/GaAs quantum dot lasers grown on silicon. Journal of Lightwave Technology. 2019;**38**:240-248

[78] Jung D, Zhang Z, Norman J, Herrick R, Kennedy M, Patel P, et al. Highly reliable low-threshold InAs

quantum dot lasers on on-axis (001) Si with 87% injection efficiency. ACS Photonics. 2018;**5**(3):1094-1100

[79] Shang C, Begley MR, Gianola DS, Bowers JE. Crack propagation in low dislocation density quantum dot lasers epitaxially grown on Si. APL Materials. 2022;**10**(1):011114

[80] Ackaert A, Buydens L, Lootens D, Van Daele P, Demeester P. Crack formation and thermal stress relaxation of GaAs on Si growth by metalorganic vapor phase epitaxy. Applied Physics Letters. 1989;**55**(21): 2187-2189

[81] Yang VK, Groenert M, Leitz CW, Pitera AJ, Currie MT, Fitzgerald EA. Crack formation in GaAs heteroepitaxial films on Si and SiGe virtual substrates. Journal of Applied Physics. 2003;**93**(7): 3859-3865

[82] Yang J, Liu Z, Jurczak P, Tang M, Li K, Pan S, et al. All-MBE grown InAs/GaAs quantum dot lasers with thin Ge buffer layer on Si substrates. Journal of Physics D: Applied Physics. 2020;**54**(3): 035103

[83] Dutta PS. Bulk crystal growth of ternary III–V semiconductors. In: Springer Handbook of Crystal Growth. Berlin Heidelberg: Springer; 2010. pp. 281-325

[84] Huang H, Ren X, Lv J, Wang Q, Song H, Cai S, et al. Crack-free GaAs epitaxy on Si by using midpatterned growth: Application to Si-based wavelength-selective photodetector. Journal of Applied Physics. 2008; **104**(11):113114

[85] Hutchinson J, Suo ZG. In: Hutchinson JW, Wu TY, editors. Advances in Applied Mechanics. Vol. 29. San Diego: Academic Press; 1992

[86] Zhao M-H, Fu R, Zhang T-Y. Multiple cracks of a thin film on a ductile substrate. MRS Online Proceedings Library. 2001;**695**:1-6

[87] Yang J, Jurczak P, Cui F, Li K, Tang M, Billiald L, et al. Thin Ge buffer layer on silicon for integration of III-V on silicon. Journal of Crystal Growth. 2019;**514**:109-113

[88] Chen Z, Segev M. Highlighting photonics: Looking into the next decade. eLight. 2021;**1**(1):2

[89] Li N, Chen G, Ng DK, Lim LW, Xue J, Ho CP, et al. Integrated lasers on silicon at communication wavelength: A progress review. Advanced Optical Materials. 2022;**10**:2201008

[90] Yang J, Tang M, Chen S, Liu H. From past to future: On-chip laser sources for photonic integrated circuits. Light: Science & Applications. 2023; **12**(1):16

[91] Reed GT, Mashanovich G, Gardes FY, Thomson D. Silicon optical modulators. Nature Photonics. 2010; **4**(8):518

[92] Michel J, Liu J, Kimerling LC. High-performance Ge-on-Si photodetectors. Nature Photonics. 2010;**4**(8):527-534

[93] Leuthold J, Koos C, Freude W. Nonlinear silicon photonics. Nature Photonics. 2010;**4**(8):535

[94] Park J-S, Tang M, Chen S, Liu H. Monolithic III–V quantum dot lasers on silicon. Frontiers of Nanoscience. 2021; **20**:353-388

[95] Cao V, Park J-S, Tang M, Zhou T, Seeds A, Chen S, et al. Recent progress of quantum dot lasers monolithically integrated on Si platform. Frontiers in Physics. 2022;**10**:839953

[96] Wang Y, Chen S, Yu Y, Zhou L, Liu L, Yang C, et al. Monolithic quantum-dot distributed feedback laser array on silicon. Optica. 2018;**5**(5): 528-533

[97] Xiang C, Liu J, Guo J, Chang L, Wang RN, Weng W, et al. Laser soliton microcombs heterogeneously integrated on silicon. Science. 2021;**373**(6550): 99-103

[98] Zhou T, Tang M, Xiang G, Xiang B, Hark S, Martin M, et al. Continuous-wave quantum dot photonic crystal lasers grown on on-axis Si (001). Nature Communications. 2020;**11**(1):977

[99] Zhou T, Ma J, Tang M, Li H, Martin M, Baron T, et al. Monolithically integrated ultralow threshold topological corner state nanolasers on silicon. ACS Photonics. 2022;**9**(12):3824-3830

[100] Shang C, Feng K, Hughes ET, Clark A, Debnath M, Koscica R, et al. Electrically pumped quantum-dot lasers grown on 300 mm patterned Si photonic wafers. Light: Science & Applications. 2022;**11**(1):299

[101] Wei W-Q, He A, Yang B, Wang Z-H, Huang J-Z, Han D, et al. Monolithic integration of embedded III-V lasers on SOI. Light: Science & Applications. 2023; **12**(1):84

Chapter 4

Molecular Beam Epitaxy of Si, Ge, and Sn and Their Compounds

Daniel Schwarz, Michael Oehme and Erich Kasper

Abstract

In the past decade, the increasing need for high-performance micro- and nanoelectronics has driven the research on group IV heterostructure devices, which utilize quantum effects as dominant working principle. The compound semiconductor SiGeSn has presented itself as promising material system for group IV heterostructures due to its unique properties. Prominent applications range from the Si-integrated laser to tunneling field effect transistors for the next complementary metal oxide semiconductor generations. However, the epitaxy of heterostructures requires atomic sharp material transitions as well as high crystal quality, conditions where molecular beam epitaxy is the method of choice since it can take place beyond the thermodynamic equilibrium. Besides the numerous opportunities, the molecular beam epitaxy of SiGeSn poses various challenges, like the limited solid solubility of Sn in Si and Ge and the segregation of Sn. In this chapter, we discuss the molecular beam epitaxy of SiGeSn at ultra-low temperatures to suppress these effects.

Keywords: molecular beam epitaxy, group IV compounds, silicon, germanium, tin, silicon-germanium-tin, heteroepitaxy

1. Introduction

Since its inception in the 1950s, the microelectronics market has grown to one of the world's most important industries. The upcoming end of Moore's law and the steady need for faster and higher-performing devices have driven the research on novel device concepts. Particularly for devices that functionalize quantum effects and heterostructures, atomic sharp structures in the nanometer regime become more and more important and attract the interest of research.

In this context, molecular beam epitaxy (MBE) is a beloved technique since it allows atomically precise deposition control as well as the process procedure beyond the thermodynamic equilibrium. Although MBE is not the best-suited tool for mass production, its uniqueness makes MBE the method of choice for research applications.

Due to the highly developed Si-based complementary metal oxide semiconductor (CMOS) technology, the MBE of Si and other group IV compounds has become more and more important in the last few decades. Possible research topics are the Si-integrated laser [1–3], the last missing key component for optical on-chip communication, as well as novel transistor concepts for future CMOS generations [4–6].

Although the beginnings of Si-MBE reached into the late 1970s and early 1980s [7–10], its number of possible applications increased rapidly since the first synthesis of Sn-containing group IV compounds [11–15]. However, generally, the epitaxy of Sn-containing group IV compounds has proven itself quite challenging due to undesirable effects, like the temperature and strain intensified segregation of Sn [16, 17], the limited solid solubility of Sn in Ge and Si of less than 1% [18, 19], and the low melting point of Sn of 231.9°C [20]. Altogether, this reasons the necessity for the execution of the MBE of group IV compounds at ultra-low substrate temperatures in the range of approximately 160°C.

In this chapter, we not only lay the basics but also discuss the particularities of the MBE of group IV compounds. For this, Section 2 deals with the physical processes on the substrate and undesired effects during the MBE process as well as the various growth modes of heteroepitaxy. Afterwards, Section 3 presents the requirements for the basic components of a group IV MBE system. Finally, Section 4 highlights some of the most important insights of the MBE of SiGeSn.

2. Basics of molecular beam epitaxy

In a typical MBE experiment, several molecular or atomic beams are directed toward a well-oriented single crystalline substrate with beam flux densities J in the order of 10^{15} cm^{-2} s^{-1} for matrix elements (group IV elements in our case) and lower 10^{12} cm^{-2} s^{-1} for doping elements (group III for p-doping, group V for n-doping). The technical solutions for beam sources are described in the next section.

The molecules or atoms on a beam have a velocity v_L of several hundred to thousand meters per second. That velocity defines two time scales. A variation of the beam intensity (on/off) is transferred to the substrate, which is typically in a distance of 10 to 50 cm, within a millisecond. In most cases, the transfer of beam on/off is defined by the technical design of the beam shutter construction. However, beam on/off switching is obtained within subnanometer and nanometer dimensions of growth, routinely. This exact control of beam fluxes makes MBE a valued method for research on nanometer heterostructures.

Another important time scale is defined by the length of interaction between the arriving beam atoms and the growing layer. Very close to the surface (nanometers or less), the atoms are attracted and accelerated toward the surface. The time scale for this interaction is in the picosecond (10^{-12} s) range.

Elastic scattering would result in a reflection of the atomic beam. The transfer of energy, momentum, and angular momentum to phonons within the picosecond time frame would result in adsorption of the atom on the surface. For single atoms with strong bonding, as in the case of group IV epitaxy, the sticking probability is near unity. This is important for good control of alloy mixing and doping with calibrated beam flux densities.

The thermal velocities v_B are given by mean energy considerations as

$$v_B^2 = \frac{3k_B}{M_P} \cdot \frac{T_B}{M} \tag{1}$$

with the constants, the Boltzmann constant $k_B = 1.38 \cdot 10^{-23}$J/K and the proton mass $M_P = 1.67 \cdot 10^{-27}$ kg and the variables, the beam source temperature T_B and the atomic weight M. For a Si beam, one will get $v_B = 1300$ m/s with $T_B = 2000$ K and $M = 28$.

2.1 Sticking, adsorption, and desorption of atoms

The sticking of atoms from group IV elements on a diamond-type lattice with strong bonds is unity. This is proven for Si [21] but can reasonably also be assumed for other group IV systems. The same conclusion is valid for the doping atoms from groups III and V, which are strongly bonded on lattice sites. Complete sticking allows the realization of complex designs with heterostructures and doping transitions using calibrated beam fluxes.

The arriving atom sticks on the surface as an isolated adatom which is weaker bonded to the crystal than a bulk atom.

$$W = W_{\mathrm{ad}} + W_S \tag{2}$$

The evaporation energy W of a bulk material is composed of the adsorption energy W_{ad} and the additional surface energy W_S, which is gained from the incorporation of an adatom on the kink of a surface step, as seen in the left part of **Figure 1**. The value of W can be extracted from the temperature-dependent vapor pressure of the matrix element. At equilibrium, the incoming flux of the vapor equals the desorbing flux density F_{eq} of the epitaxial surface. The typical value of W amounts to several electronvolts per atom (Si: $W = 4.55$ eV/atom).

The adsorption energy W_{ad} depends on the surface orientation and on the reconstruction of the surface. For model calculations, often a value of $2/3W$ is assumed. The impinging flux F_S of atoms is usually much higher than the desorbing flux because MBE conditions are preferably in the lower growth temperature regime. That means the supersaturation σ is high. We define here the supersaturation as

$$\sigma = \frac{F_S}{F_{\mathrm{eq}}} = 1 \tag{3}$$

Sometimes, the value of $\ln \sigma$ is also called supersaturation. It must be considered that supersaturation under growth conditions causes an increase in the concentration of adatoms from the equilibrium value n_{s0} to an n_s, which is dependent on the distance to the next step.

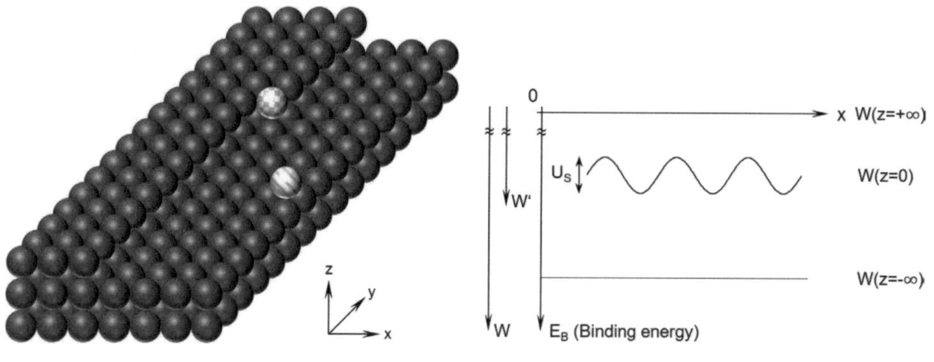

Figure 1.
Adsorption and incorporation of an atomic beam. Left: Schematic view of a substrate surface with an adatom (striped), a surface step, and an incorporated adatom at a kink side of an atomic step (checked). Right: Binding energy diagram for the movement of an adatom in x-direction with the adsorption energy W_{ad}, the bulk crystal binding energy W, and the surface diffusion barrier U_S.

Steps on the surface, mainly of mono-atomic or bi-atomic height h, stem from three sources.

- Slight surface misorientation of the nominal surface with low index planes (Miller index: (001) planes perpendicular to the cubic axes, (111) planes perpendicular to the cubic diagonals). The preferred surface orientation for Si substrates is (001) because the interface quality between dielectrics and semiconductors is the best. Commercially available substrates typically have a misorientation angle i below $i < 0.25° \left(\text{arc } i \leq 4 \cdot 10^{-3}\right)$

$$\text{arc } i = \frac{h}{L} \tag{4}$$

  with the atomic step height h and the length of the terrace to the next step L. With $h = 0.14$ nm, the mono-atomic step height on Si(001), one obtains terrace lengths of $L > 35$ nm. That means commercial substrates already deliver a dense sequence of steps. The microscopic picture of the macroscopic growth in vertical direction is, therefore, a lateral movement of step trains. The steps move in the downward direction by the repeated incorporation of adatoms to the kink position on steps.

- Dislocations generate a step at their point of intersection with the surface. These steps are pinned to the dislocation position. Under the supersaturation of adatoms, they wind up as spirals. These sources are available in substrates of lower quality than Si or in virtual substrates formed by relaxed layers on Si, for example, Ge virtual substrates. Burton-Cabrera-Frank [22] developed their early atomic theory of growth from step movement under this assumption.

- Two-dimensional nucleation generates additional steps. This mechanism operates if the concentration of adatoms is very high. This concentration depends on the supersaturation of the incoming flux and the diffusion of adatoms toward the steps. The diffusion of adatoms is described by a diffusion coefficient D_S which results from a Brownian motion across an energy barrier U_S (see right part of **Figure 1**) between the lattice sites for adatoms. One should consider the surface as populated with very mobile adatoms. Epitaxy by MBE is, therefore, possible at low temperatures when bulk atoms are frozen into their respective positions. However, adatoms become less mobile at lower temperatures. Finally, the adatom flux to preexisting steps does not counterbalance the incoming atom flux. The resulting increasing adatom density promotes nucleation processes, as seen in **Figure 2**, mainly by reducing the needed size for the so-called critical nucleus.

The critical nucleus forms by a dynamic process counterbalancing the incoming adatom flux, which is proportional to the adatom density, and the leaving adatom flux which is dependent on the nucleus size. In effect, this means an increase of steps when the surface adatom supersaturation σ_S surpasses a value necessary for the formation of critical nuclei.

$$\sigma_S = \frac{n_S}{n_{S0}} - 1 \tag{5}$$

There is an important difference of these nuclei steps and steps from misorientation. The step trains from misorientation follow the inclination direction.

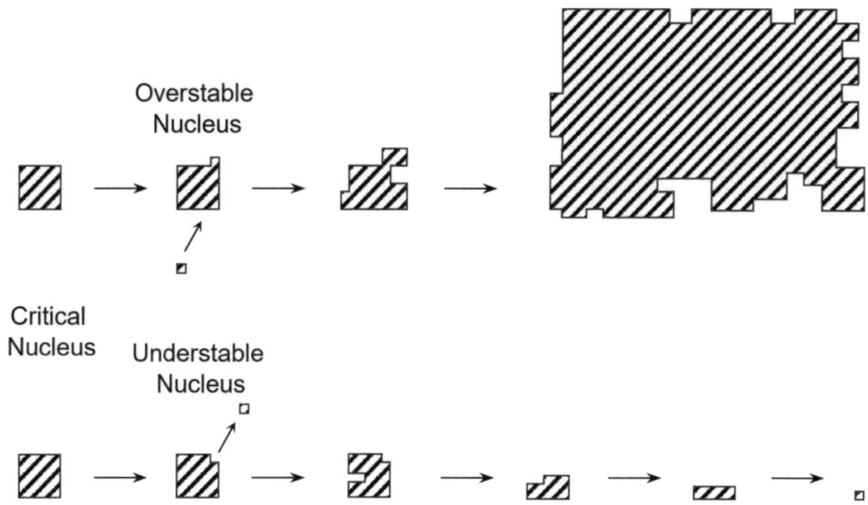

Figure 2.
Schematic presentation of the two-dimensional nucleation. Top: Development of a stable nucleus. Bottom: Decay of a critical nucleus.

The steps from adjacent nuclei move forward and become extinct when they meet. For each mono-atomic step, a new nuclei must be formed. The caused periodic modulation of the step density may be observed with sensitive methods like RHEED (reflection high energy electron diffraction).

2.2 Growth modes of strained heteroepitaxy

The macroscopic growth mode models, as seen in **Figure 3**, predict two-dimensional (2D) growth (Frank v. d. Merve), three-dimensional (3D) growth (Volmer-Weber), or mixed 2D/3D growth (Stranski-Krastanov) from a balance of surface energies of the substrate E_S and the film E_f and the interface energy E_i densities

$$E_S - E_i = E_f \cdot \cos\Theta \tag{6}$$

The inclination angle $\Theta \geq 0$ of a 3D island is given by Eq. (6). Solutions with $\cos\Theta \geq 1$ deliver 2D growth. Mixed 2D/3D growth mode is only possible for a thickness-dependent E_i term

Microscopic theories explain the thickness dependence of the interface energy by contributions from the elastic film strain and from the interface misfit-dislocation network. **Table 1** gives the lattice constants of the diamond-type lattice cell for the group IV elements C, Si, Ge, and Sn.

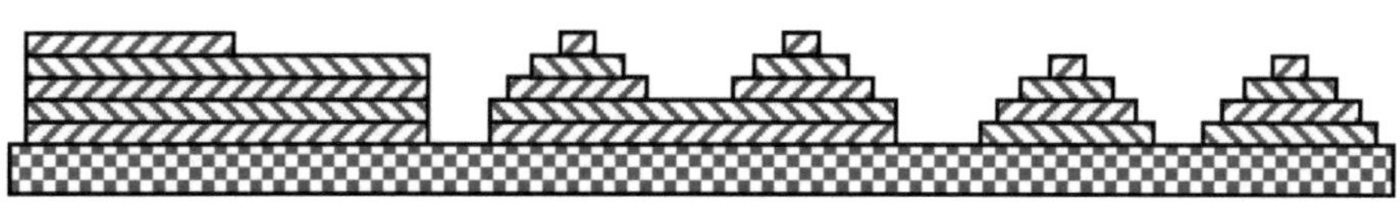

Figure 3.
Monocrystalline growth models on a crystalline substrate (checked). Left: Two-dimensional (2D) growth according to Frank van der Merve. Middle: Three-dimensional (3D) growth according to Volmer-Weber. Right: Mixed two-three-dimensional growth according to Stranski-Krastanov.

	C	Si	Ge	Sn
a_0 (Å)	3.56683 [23]	5.43102 [23]	5.64613 [23]	6.4912 [24]

Table 1.
Lattice constants of the group IV elements C, Si, Ge, and Sn.

For the group IV compound SiGeSn, the lattice constant $a_{0,\mathrm{SiGeSn}}$ depends on the compound composition and follows the relationship in Eq. (7):

$$a_{0,\mathrm{SiGeSn}} = c_{\mathrm{Si}} \cdot a_{\mathrm{Si}} + c_{\mathrm{Ge}} \cdot a_{\mathrm{Ge}} + c_{\mathrm{Sn}} \cdot a_{\mathrm{Sn}} \tag{7}$$

In heteroepitaxy, two materials, hence the film and the substrate, with different electrical and mechanical properties but the same crystal structure, are grown on each other. Therefore, the lattice constant of the grown film a_f usually differs from the lattice constant of the substrate a_S. The relative lattice constant difference, according to Eq. (8), is called lattice mismatch f:

$$f = \frac{a_f - a_S}{a_S} \tag{8}$$

Due to the lattice mismatch, three possible cases, as shown in **Figure 4**, are conceivable: For pseudomorphic, also known as coherent, growth, the atomic rows of the film fit exactly to the underlying rows of the substrate. In other words, the in-plane lattice constant of the film $a_{\parallel f}$ adapts to the lattice constant of the substrate a_S (see left part of **Figure 4**). However, due to the different lattice constants of the two materials, this can only take place under the formation of elastic strain of the film with the in-plane strain $\varepsilon_{\parallel} = -f$.

The elastic strain energy of a pseudomorphic film is proportional to the film thickness. This pseudomorphic state holds only up to a critical thickness t_C of the film. Above this critical thickness for strained films, dislocations are nucleated and bend to the interface for a partial relief of strain. Nanometer films and structures may be strained to much higher values, for example 1% strain in 10 nm thickness, than bulk material. This high tolerance of high strain values in nanometer structures is the base of strain engineering in electronics and photonics. Films with a thickness above the critical thickness, but which exhibit residual strain, are referred to as partially relaxed (see middle part of **Figure 4**).

In the case of complete relief of strain, the resulting film is called completely strain relaxed. The corresponding growth mode is thus called strain-relaxed growth.

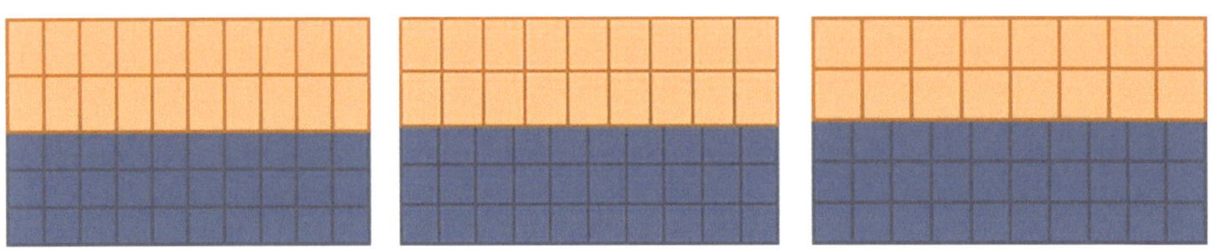

Figure 4.
Comparison of the growth modes of heteroepitaxy with the grown film (brown) on the substrate (blue) left: Pseudomorphic growth. Middle: Partial-relaxed growth. Right: Strain-relaxed growth.

A special case of heteroepitaxy can be performed with SiGeSn. The mixing of Si and Sn with Ge allows a decrease of the lattice constant with Si and an increase with Sn. Therefore, SiGeSn allows the decoupling of its electrical and mechanical properties from its lattice constant. Consequently, SiGeSn can be grown lattice-matched on Ge and even on GeSn. In order to achieve this so-called lattice-matching on Ge, a constant ratio of the Si and Sn concentration, according to Eq. (9), has to be fulfilled:

$$\frac{c_{\mathrm{Si}}}{c_{Sn}} = \frac{a_{\mathrm{Ge}} - a_{\mathrm{Sn}}}{a_{\mathrm{Si}} - a_{\mathrm{Ge}}} = 3.67 \tag{9}$$

2.3 Limits of single crystalline growth

Considering a strongly magnified picture of the epitaxy process reveals that approx. 10^{15} atoms/(cm^{2}s) are impinging on the surface. In the MBE regime without desorption, the growth rate R is determined by Eq. (10):

$$R = F_S \cdot \Omega \tag{10}$$

with the beam flux density j and the atomic volume $\Omega = 2 \cdot 10^{-23}$ cm^3 for Si.

The growth rate equals a coverage of approx. one monolayer (ML) per second. This means each atom out of the 10^{15} atoms/(cm^{2}s) have to find their position on a lattice site within roughly 1 s; otherwise, the next monolayer flux would cover and fix them on unwanted sites, producing interstitials and vacancies. Exactly that is happening at very low growth temperatures of approx. 20–100°C, resulting in a growth start with single crystalline layers of high crystal defect density and subsequently an alteration to amorphous material after a critical thickness of crystalline growth at low temperatures. It has to be considered that this low-temperature growth *critical thickness* is not the same as that of the coherent growth of mismatched structures. This defect-rich growth regime is not applicable for working devices but is useful as a nucleation starter for dislocations [25] and amorphous structures.

2.4 Segregation

Through the usage of MBE, the influences of the volume diffusion of doping or alloy atoms within the Si or SiGeSn crystal can be remarkably reduced. However, the effect of surface segregation, which results in a broadening or smearing of doping profiles, can also be observed in these epitaxial layers. In order to suppress this effect, special epitaxial strategies are needed. The basic mechanism of surface segregation is an exchange of atoms from subsurface states to surface states due to the energetically favorable surface adatom position (see **Figure 5**). The segregation is quantified by the segregation length Δs [9], which is the ratio of the impurity surface concentration n_S and the bulk concentration n_B. The segregation length Δs, thus defined, can be converted into the dimensionless segregation coefficient S by dividing it by the thickness of a monolayer. Segregation leads, therefore, to an increasing impurity surface concentration n_S, which saturates at an equilibrium where the number of segregating adatoms equals the number of incorporating adatoms. Therefore, the resulting doping profile shows an exponential increase after the beginning of the co-evaporation of dopant and lattice atoms. Consequently, the segregation length Δs describes the length in the growth direction from the beginning of co-evaporation to reaching equilibrium.

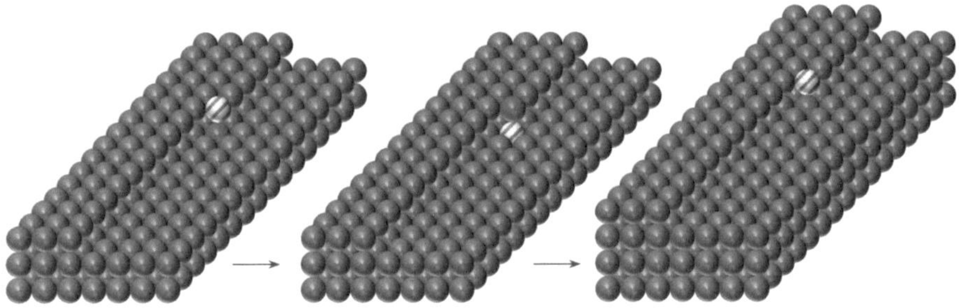

Figure 5.
Left: Dopant atom (blue) in subsurface state. Middle: Dopant atom swaps its place with a surface crystal adatom (gray striped). Right: Further overgrowth leads to the same situation as before.

The segregation length is typically in the order of a few nanometers but can, outside of the optimum growth parameter window, increase to several tenth to hundred nanometers.

Generally, the perfect abruptness of transitions may be obtained by a growth interruption, in which the equilibrium adatom concentration of the doping or alloy element is adjusted. These techniques are called prebuild-up for ramp-up and flash-off for an abrupt ramp-down.

The surface segregation of Sb in Si and Ge is strongly dependent on the growth temperature [9, 26–28]. In order to achieve an atomic sharp doping profile, the growth strategy, which is known as prebuild-up, can be used. For this, the adatom concentration required for equilibrium, which is typically less than a monolayer, is evaporated onto the surface before the co-evaporation begins. At the beginning of the subsequent co-evaporation, the number of incorporated adatoms per monolayer is already at the desired value.

The biggest disadvantage of the prebuild-up technique to counteract surface segregation is the remaining surface concentration of segregated atoms. Following the segregation mechanism, the surface concentration decreases linearly subsequently to the co-evaporation of dopant and matrix atoms, which leads in turn to a nonideal doping profile. Therefore, special methods are needed to get rid of the segregated surface concentration such as the flash-off method. Here, a growth interruption is used for an increase of the substrate temperature to a value where the remaining segregated atoms are being desorbed from the epitaxial surface. However, depending on the element to be desorbed, the necessary temperature for this can be relatively high and in an unfavorable range for the already grown crystal. In order to avoid the disadvantages of high temperature flash-off, the "freeze out" method may be used by which growth rate and temperature are reduced to form a delta-doping structure from the adlayer. This is specially of advantage in device structures with highly doped regions in contacts, in buried layers, and in modulation-doped quantum wells.

Besides the described prebuild-up technique, other methods can be used to suppress or counteract surface segregation. A possibility is the doping by secondary implantation (DSI) [29, 30]. In the example of Si, which is being evaporated using an electron beam evaporator (see Section 3.1.2), the molecular beam contains Si atoms as well as Si^+ ions. These Si^+ ions can then be accelerated toward the substrate using an electric field. Once these accelerated ions hit the substrate surface, they can drive segregated atoms further into the crystal and, therefore, counteract the segregation. Furthermore, special effusion cells are conceivable in which the evaporated atoms are ionized in an additional stage, whereby the dopants themselves can be accelerated toward the substrate surface [31, 32].

The presence of a third species of adatoms also influences the segregation length because the species with higher binding energy occupies the crystal positions faster. We call the third species a surfactant if it is added only to reduce the segregation length of the second species.

Particularly in the MBE of group IV compounds such as SiGeSn, segregation has a huge influence on the incorporation of Sn [16, 33]. In this case, segregation not only leads to a smearing out of the alloy composition but also to a disturbance of the epitaxy process. The consequences of Sn segregation range from an increased point defect concentration within the crystal over the formation of Sn precipitates on the surface and finally to the complete breakdown of the epitaxial process (see Section 4). Consequently, many of the previously described methods to counteract the surface segregation are not applicable for the MBE of SiGeSn which in turn justifies the necessity of ultra-low-temperature MBE of these group IV compounds.

3. Molecular beam epitaxy of group IV compounds and low-dimensional structures

3.1 Molecular beam sources

3.1.1 Theory of molecular beam sources

Different types of molecular beam sources are used in MBE systems depending on the physical properties of the material to be evaporated or sublimated. The equilibrium vapor pressure p_0 as a function of the absolute temperature T of the vaporization material plays an important role. From this dependency, the particle flux F_0 emitted by a surface is calculated with Eq. (11) as follows:

$$F_0 = p_0 \sqrt{\frac{N_A}{2\pi M_r k_B T}} \tag{11}$$

with the surface temperature T, the molecular weight M_r of the considered element or molecule, the Avogadro constant N_A, and the Boltzmann constant k_B. From the evaporation rate Γ_0 at the surface of a melt, the area-specific particle flux can be calculated, which vaporizes an area A at a distance r and at an angle ϑ. The flux F_S arriving there is inversely proportional to the square of the distance r and has the same characteristics as the evaporation rate to a first approximation. At some distance, the effusion cell acts like a point source of flux with a cosine distribution. The particle flux F_S impinging on the substrate is then given with Eq. (12) as follows:

$$F_S = \frac{\Gamma_0}{r^2} \cos\vartheta = \frac{F_0 A}{r^2} \cos\vartheta \tag{12}$$

The required working temperatures can be determined with these formulas. For this purpose, the vapor pressures as a function of the temperature for the corresponding elements are taken from the literature. It is additionally assumed that the typical evaporation area is 1 cm^2, and the distance between the evaporation surface and the substrate is $r = 30$ cm. Furthermore, perpendicular incidence is assumed ($\vartheta = 0$).

The left graph in **Figure 6** shows the results of the calculations for the matrix materials of group IV. Therefore, the necessary temperature range to achieve the required growth rates for the respective molecular beam sources can be seen in this diagram. The growth rate R corresponds to the right scale on the example of a Si(100) surface.

For a Si molecular beam source, a working temperature range between $1460°C \leq T \leq 1820°C$ results in growth rates of $0.1Å/s \leq R \leq 10Å/s$. Using the melting point of Si ($T_{\text{melt,Si}} = 1410°C$), it follows that Si is evaporated instead of sublimated, which means that a crucible must be used.

In case of the growth of carbon, the working range of a molecular beam source is between $2270°C \leq T \leq 2630°C$ for growth rates of $0.1Å/s \leq R \leq 10Å/s$. However, since the melting point of carbon ($T_{\text{melt,C}} = 3550°C$) is significantly higher, carbon is only sublimated under MBE conditions. Therefore, carbon can be used directly as a filament material.

The right-hand graph of **Figure 6** shows the particle fluxes on the substrate as a function of the surface temperature of the material to be evaporated for typical group IV doping materials. In Si-based technology, only the element B is used for p-type doping. For a large variation of the doping concentration between $10^{16}\text{ cm}^{-3} \leq N_A \leq 10^{20}\text{ cm}^{-3}$, the resulting typical working temperature of a B source in the range of $1200°C \leq T \leq 1600°C$, as it can be read in the vapor pressure curve. Since the melting point of B is $T_{\text{melt,B}} = 2076°C$, B belongs to the sublimated materials under MBE conditions.

The typical n-type dopants in Si technology are P and As, although Sb is also used. The respective vapor pressure curves are also shown in the right graph of **Figure 6**. The typical temperatures for the evaporation of the element P for doping concentrations between $10^{16}\text{ cm}^{-3} \leq N_D \leq 10^{20}\text{ cm}^{-3}$ are in this regard in the range of $33°C \leq T \leq 97°C$. This is very problematic in an ultra-high vacuum (UHV) chamber, since an effusion cell coats not only the substrate but also the rest of the chamber, specifically the chamber walls. However, these P deposits evaporate again at low temperatures, resulting in a continuous flow of P throughout the growth chamber. The result is a tremendous background doping in the epitaxial layer.

Things are looking a little better for the next group V element, As. The typical temperature range here is between $81°C \leq T \leq 172°C$. For the conditioning of the

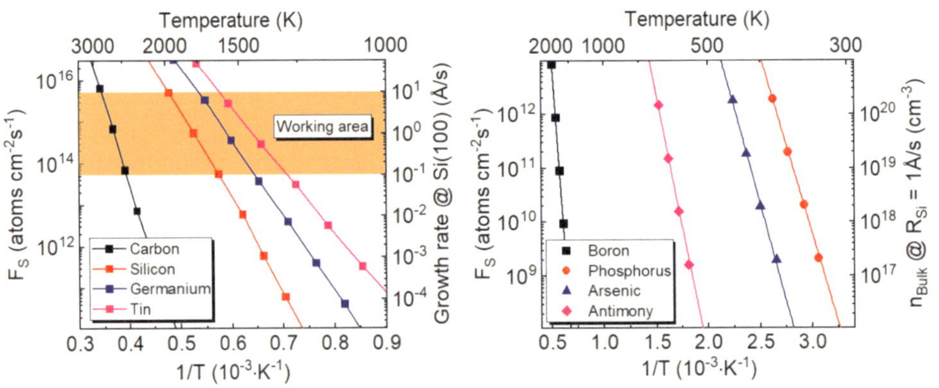

Figure 6.
Particle flux on the substrate as a function of the inverse surface temperature $1/T$ *of the evaporation material. Left: Typical group IV matrix materials. Right: Typical doping materials for Si-based devices. (data source: [34]).*

UHV, the entire chamber is heated at $T \geq 200°C$. In the case of residual As in the chamber, it would be distributed evenly in the chamber during this conditioning, leading to a similarly increased background doping.

Consequently, the group V element Sb is typically used as n-type dopant in a Si-MBE system. The resulting working range of an Sb beam source is between $240°C \leq T \leq 380°C$. Since the melting point of Sb is $T_{melt,Sb} = 630°C$, Sb is being sublimated under MBE condition.

The choice of the appropriate molecular beam source is based on the vapor pressure curve and the melting point of the material to be evaporated. The possibilities include electron beam evaporators (EBE), effusion cells, and high-temperature sublimation cells.

3.1.2 Electron beam evaporators

Although any material can be evaporated using an EBE, their technical realization is very challenging. A schematic drawing of the main components and the functionality of a Si EBE is shown in the left part of **Figure 7**.

A W filament emits electrons which are accelerated in an electric field with typical energies between 8 and 12 keV. Since metallic impurities in the Si melt are undesirable, the electrons, emitted from the filament, are thereupon deflected at an angle of 270° using a magnetostatic field, which does not manipulate the also emitted W atoms. Due to the high reactivity of molten Si, it would attack any crucible material, which in turn leads to an excessive contamination of the melt and the molecular beam with the crucible material itself. Therefore, Si acts in an EBE as its own crucible material. For this purpose, a high-purity Si ingot is used, which is placed in a water-cooled Cu crucible. The electron beam melts the Si only partially on the surface. By using an additional electromagnetic field, the electron beam can be scanned over the entire crucible to adjust the molten area. Due to the thermal properties of Si, the molten area is limited to the area directly heated by the electron beam.

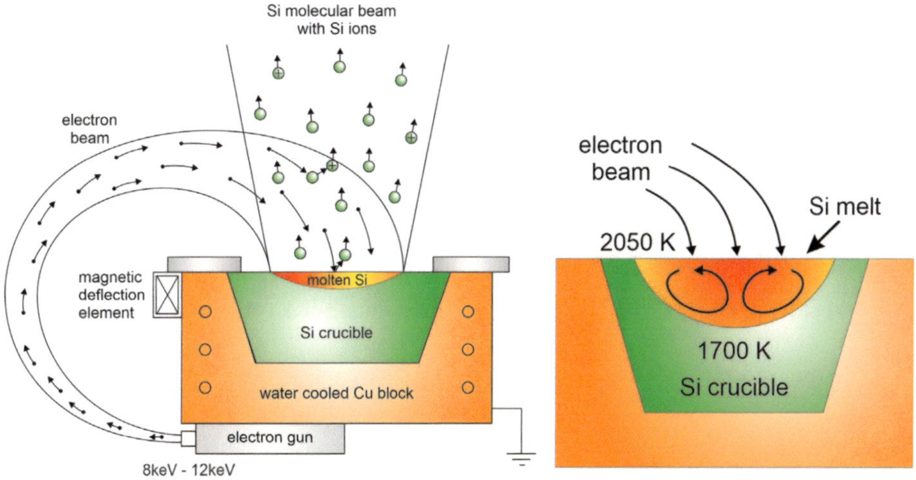

Figure 7.
Left: Schematic drawing of the main components and the functionality of an electron beam evaporator (EBE) for the evaporation of Si. Right: Marangoni convection caused turbulence in the Si melt.

The EBE is controlled using a quadrupole mass spectrometer (QMS), which directly measures the intensity of the Si isotope with the mass number 30 in the Si molecular beam. The measurement signal is proportional to the flux or the growth rate on the substrate surface and can be used as a control parameter. A problem of heating with electrons is the Marangoni convection in the Si melt, as seen in the right part of **Figure** 7.

Since there are subareas of different temperatures with different surface tensions in the molten Si, turbulent flows arise in the melt, leading to high flux fluctuations. Although this effect can be compensated by the regulation using a QMS, the flux of an EBE underlies higher fluctuations compared to an effusion cell. However, the biggest advantage of an EBE in comparison to effusion cells is the high rate of change of the desired Si flux. The disadvantage, on the other hand, is the already mentioned fluctuation of the Si rate, which is in the range of approx. 10%.

3.1.3 Effusion cells

In an effusion cell, the crucible is heated from the outside and from below by thermal radiation. For this purpose, the crucible is surrounded by an electric heater. The production of uniform layers requires a constant flux over time. The flux from an effusion cell is determined indirectly based on the cell temperature and can be regulated by the applied heating power. The task of controlling a cell is to keep the cell temperature as constant as possible at a specified value, or to reach a new nominal temperature as quickly as possible. The regulation of the temperature of an effusion cell by the heating power must be very precise since the evaporated flux increases exponentially with the temperature.

The thermal properties of an effusion cell are the static and dynamic relationships between the conversion of electrical energy into heat, the propagation and dissipation of the thermal energy, and its effect on the evaporated flux. The material flux can only be controlled precisely if its reaction to the electrical power input is known. Since the heating element is located inside the effusion cell, the electrical heating power is initially transferred to the cell. If one neglects delays due to heat capacities, a power balance can be drawn up at any point in time: The sum of all power losses corresponds to the electrical power input. Thermal energy can be transported between two bodies by the following three mechanisms:

1. Thermal radiation

2. Conduction

3. Convection

This results in the following heat loss mechanisms for a typical effusion cell:

1. Thermal radiation through opening the shutter

2. Thermal radiation through the shield

3. Heat conduction *via* fortifications

4. Evaporation energy and kinetic energy of the molecular beam

The ratio of the parts in the total power depends on the temperature of the cell and, thus on the induced electrical power. For the most thermally stable operation possible, the power loss must be minimized.

A schematic drawing of the structure of an effusion cell can be seen in the left part of **Figure 8**. The evaporating material is placed in a pyrolytic BN crucible (PBN). This crucible is heated using a meander-shaped graphite heater. Side and bottom shields made of Ta minimize the thermal losses of the effusion cell. A water-carrying cooling shroud is used for a further reduction of the thermal radiation into the MBE chamber. The temperature of the crucible is measured using a thermocouple, which has to be calibrated to the actual temperature of the melt. The molecular flux can be switched on or off with a shutter.

The illustration on the right part of **Figure 8** shows a technical realization of a Ge effusion cell. The effusion cell is built on a flange with a diameter of 150 mm and has a crucible with a volume of 100 cm^3. The built-in thermocouple of an effusion cell is used to monitor and control it. Besides that, it is also possible to directly measure and control the atomic or molecular flux using a QMS. The advantage of an effusion cell is the very stable molecular flux. However, due to the good thermal isolation of the cell, the rate of change in the molecular flux is very limited, in comparison with an EBE.

3.2 *In situ* reflectometry

The *in situ* analysis describes the measurement of various process parameters during the epitaxy process in an MBE chamber. In addition to measuring the composition of the residual gas, the actual surface temperature of the substrate, a measurement of the reflectivity of the epitaxial layers can also be carried out directly during the process. The so-called *in situ* reflectometry is interesting when the optical properties of the epitaxial layer change compared to the underlying substrate or layer

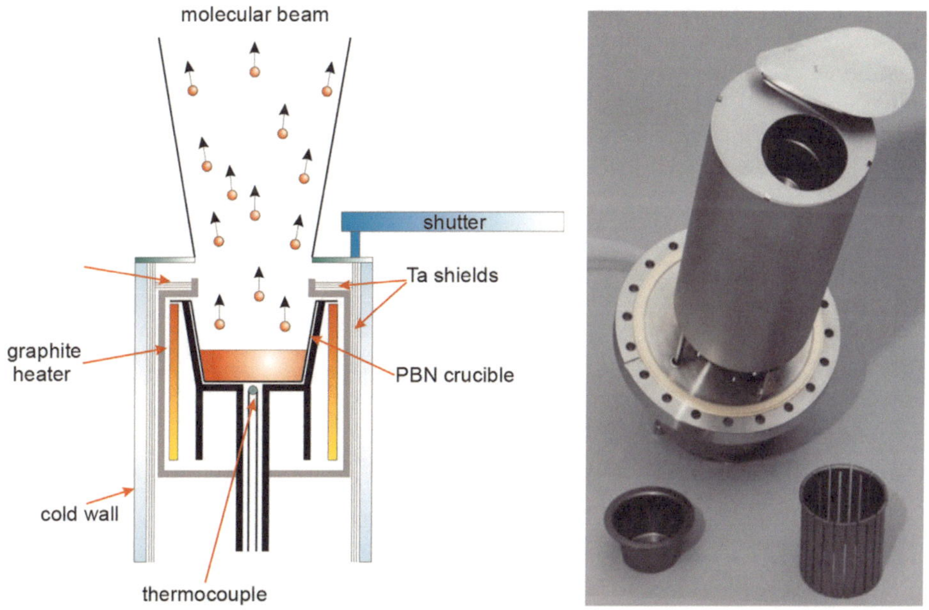

Figure 8.
Left: Schematic scheme of an effusion cell. Right: Real Ge effusion cell on a 150 mm flange. The crucible has a volume of 100 cm³. The meander-shaped heater is made of graphite.

stack. For example, a change from monocrystalline to amorphous growth can be observed. In the case of heteroepitaxial growth, the thickness of the growing layer can also be measured using *in situ* reflectometry.

The perpendicularly incident light beam is reflected both at the growth surface and at the interface between the substrate and the heterolayer, as it can be seen in the left part of **Figure 9**, which leads to the interference of the respective parts.

A typical measurement of the reflectivity during a growth process is shown in the right part of **Figure 9**. The growth typically begins with a Si buffer on the Si substrate. Since the optical properties do not change between the film and the substrate, the reflectivity measurements remain constant at both used wavelengths. At a time $t = 0$ s, the heteroepitaxial growth of a SiGe layer begins. Therefore, the reflectivity of the epitaxial layer changes from $n_{epi} = n_{sub} = n_{Si}$ to then $n_{epi} = n_{SiGe}$ immediately with increasing thickness of the SiGe layer. Therefore, the already mentioned thickness interference can be observed. Using the positions of the arising maxima and minima, the corresponding thickness d of the epitaxial SiGe layer can be calculated using Eq. (13) and Eq. (14), respectively.

For $n_{epi} > n_{sub}$ the following Eq. (13) applies for the maxima:

$$n_{epi} \cdot d = \frac{\lambda}{4} \cdot (2m + 1) \tag{13}$$

and Eq. (14) applies for the minima:

$$n_{epi} \cdot d = \frac{\lambda}{2} \cdot m \text{ with } m \in \mathbb{N} \tag{14}$$

with the wavelength λ of the used light.

Besides that, a weakening of the oscillations for the wavelength of 670 nm is measured. This is due to the absorption behavior of the SiGe layer. Consequently, the absorption behavior of the epitaxial-grown material can be determined from these measurements. On the other hand, at a wavelength of 950 nm, the absorption is negligible.

The technical implementation of an *in situ* reflection measurement is shown in **Figure 10**. In order to enable the necessary perpendicular incidence of light, a flange

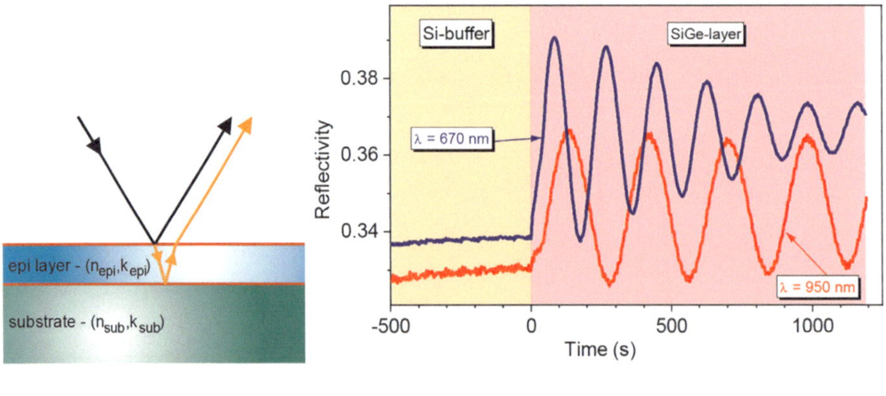

Figure 9.
Left: Schematic drawing of the reflectometry measurement principle. Right: Exemplary in situ *reflection measurements at the wavelengths of 670 and 950 nm during the growth of an Si buffer and a SiGe layer on a Si substrate.*

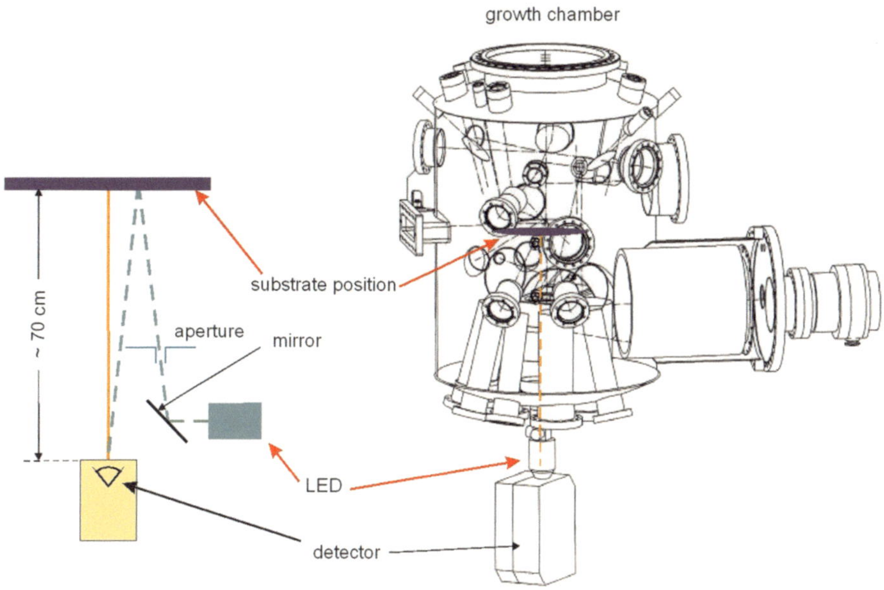

Figure 10.
Schematic drawing of the setup for in situ *measurement of the reflectivity of the growing layer. The measuring system is attached to the growth chamber in such a way that the reflectivity can be measured under vertical incidence of light.*

with a window is mounted directly under the substrate on the growth chamber. The measurement head contains several light sources at different wavelengths and a corresponding detector. For an optimal homogeneity of the layer thickness over the entire wafer, the substrate is rotated during growth. In order to compensate for slight wobbling movements, the reflectivity measurement is synchronized with the rotation.

3.3 Substrate heating and temperature measurement

Heating of the substrate surface during the epitaxy process is necessary due to various reasons. As mentioned in Section 2.1, heating provides not only enough energy for adatoms to reach their preferred destination on the crystal surface. Furthermore, heating of the substrate prevents the adsorption of undesired elements of the residual gas and, therefore, the concentration of impurities.

As described in Section 3.1, thermal energy can be transported by three different mechanisms. Due to the lack of a transport medium as well as the necessity for substrate rotation, the mechanism of choice for the transport of thermal energy under MBE conditions is thermal radiation. The typical temperature regime for the substrate heating for the epitaxy of group IV compounds is in the range of $100°C \leq T_S \leq 1200°C$. According to Planck's radiation law, the corresponding radiation is, as seen in the left part of **Figure 11**, in the wavelength range of $\lambda \geq 1100$ nm. However, the absorption of typically used Si substrates is due to the bandgap of Si of $E_g = 1.12$ eV limited to wavelengths $\lambda < 1100$ nm (see the absorption border of Si in the left part of **Figure 11**). Therefore, the heating of Si substrates in a group IV MBE system is based not on the fundamental absorption but on the free carrier absorption.

The necessary infrared radiation for the heating of the substrates is generated using a graphite meander-based electrical heater. A schematic drawing of such a heater is shown in the right part of **Figure 11**. The measurement and control of the

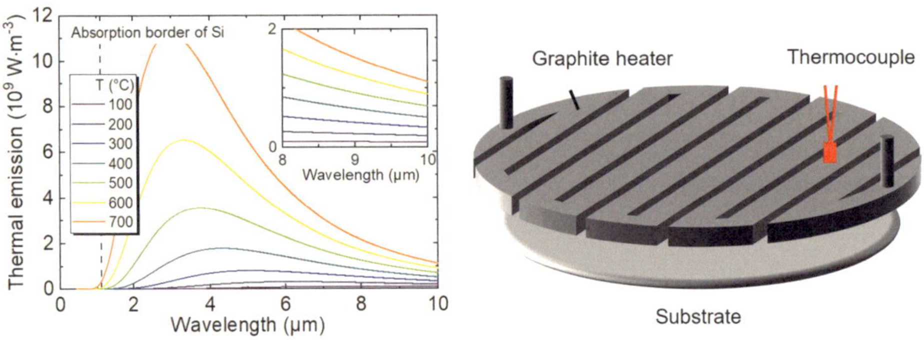

Figure 11.
Left: Thermal emission of a radiative heater according to Planck's radiation law. Right: Schematic drawing of a graphite meander-based electrical heater and the used thermocouple for measurement and control of the substrate temperature.

substrate temperature is generally performed using a thermocouple, which is located behind the graphite heater. Due to the heating *via* free carrier absorption, the measurement of the thermocouple is calibrated once to the actual substrate surface temperature using a second thermocouple mounted to the surface of a calibration wafer. Since the amount of free carriers in a semiconductor can be manipulated *via* doping, the resulting calibration is dependent on the substrate doping concentration.

Despite the necessity of substrate heating, too high substrate temperatures promote undesired, epitaxy disturbing effects, especially the already in Section 2.4 described segregation. In particular, the MBE of group IV compounds suffers from the segregation of Sn [16, 33], which causes its often-reported temperature sensitivity. Furthermore, the limited solid solubility of Sn in Ge of less than 1%, the instability of the preferred α-Sn phase with its transition to β-Sn at 13.2°C, and the occurring in-plane strain of GeSn on Ge amplify this effect. While not only hindering the incorporation of Sn, the segregation also leads to a high number of lattice defects, and, due to the increasing surface concentration of Sn, to β-Sn clusters and the subsequent epitaxial breakdown. All this together underlines the necessity to perform the epitaxy of SiGeSn at ultra-low substrate temperatures in the range of $T_S \approx 160°C$.

However, in this temperature regime, not only the substrate heater itself has an influence on the actual substrate temperature but also the thermal radiation of the molecular beam sources. In fact, their influence is, in this regime, the more dominant one and restricts the controllability of the substrate temperature on the base of thermocouple measurement. Therefore, an alternative method for the measurement of the actual substrate surface temperature is needed.

In this context, the mid-infrared (MIR) pyrometry in the wavelength range of $8\ \mu m \leq \lambda \leq 14\ \mu m$ has proven itself as a suitable solution. Possible realizations include mid-infrared cameras as well as single detector pyrometers. The left part of **Figure 12** shows a mid-infrared picture of an epitaxial GeSn surface. Besides that, the graph in the right part displays the comparison of the substrate temperature measured once with the thermocouple and the mid-infrared pyrometer in the moment of the growth beginning of an exemplary GeSn layer, more precisely, the opening of the Ge and Sn shutters. For this growth process, the heating power was set to $P_{Heat} = 0$ W to achieve the desired ultra-low substrate temperature.

Particularly remarkable is the observable increase of the substrate temperature by $\Delta T = 30$ K, which can only be explained by the thermal radiation of the molecular

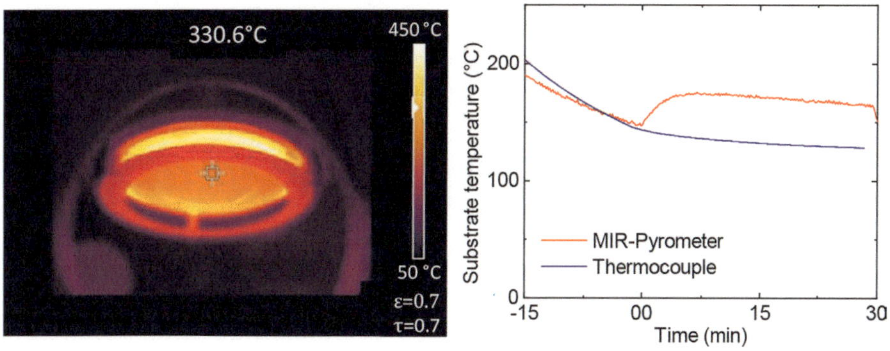

Figure 12.
Left: MIR-picture of a Si substrate during a GeSn growth process. Right: Characteristics of the substrate temperature according to the calibrated thermocouple (blue) and the MIR pyrometer (red) at the beginning of a GeSn growth process.

sources. Considering the required substrate temperature of $T_S \cong 160°C$, this increase underlines the huge impact of the source radiation on the actual substrate surface temperature.

4. Optimization of the molecular beam epitaxy of SiGeSn

4.1 Influence of the molecular flux distribution on SiGeSn epitaxy

As already mentioned in Section 2.2, the electrical and mechanical properties of SiGeSn strongly depend on its composition. Furthermore, the desired condition for lattice-matched growth of SiGeSn on Ge requires a constant ratio of Si to Sn according to Eq. (9). All this underlines the necessity for an as homogeneous as possible distribution of the involved molecular fluxes of Si, Ge, and Sn. However, as already described in Section 3.1.1, the flux F_S impinging on the substrate follows a cosine distribution of the irradiation angle. Therefore, the installation position and angle of a molecular beam source strongly influence the homogeneity of the epitaxial layer. Since only one source can be placed directly central under the substrate, the sources must be arranged in a circular and tilted manner under the substrate, which leads to an even worse distribution. This results, in turn, in the necessity of substrate rotation during growth to achieve at least circular homogeneity of the epitaxial layers. Altogether, the result is a radial-dependent change in the alloy composition, as seen in the exemplary flux distribution on a 4″ substrate, as shown in the left part of **Figure 13**.

As can be seen, the exemplary MBE system shows a divergence of the Si and the Sn flux. Particularly for SiGeSn, this means that the condition for lattice-matching is, in the best case, only fulfilled at one radius. The resulting SiGeSn film is, therefore, not completely strain-relaxed but only strain-reduced. The distribution of the residual strain $\varepsilon_{\parallel,\mathrm{SiGeSn}}$ of an exemplary SiGeSn layer with the lattice-matching fulfilled in the substrate center is shown in the right part of **Figure 13**.

4.2 Influence of the growth temperature on the segregation of Sn

It was already previously mentioned that the admixture of Sn to group IV compounds complicates their epitaxy. The main reason for this is the segregation of Sn,

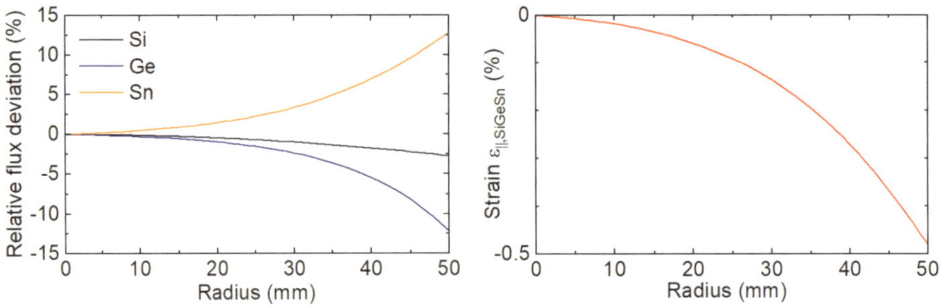

Figure 13.
Left: Relative flux deviation of Si, Ge, and Sn for an exemplary MBE system on a 4" substrate. Right: Residual strain distribution of a SiGeSn layer with the lattice-matching fulfilled in the center of the substrate.

which is, as already explained in Section 2.4, strongly dependent on the growth temperature. It was already reported that Sn segregation seems to be intensified when in-plane strain of the grown film comes into play, as it is the case for Sn-rich GeSn (SiGeSn with $c_{Si} = 0; c_{Sn} > 8\%$) [17].

However, by the admixture of Si to GeSn, thus the epitaxy of SiGeSn, it has been shown that lattice-matched film growth on Ge can be achieved by fulfilling the condition in Eq. (9). Since the in-plane strain is then almost zero, the segregation of Sn is therefore expected to be much lower as for the epitaxy of Sn-rich GeSn. Therefore, higher substrate temperatures should be possible for the epitaxy of lattice-matched SiGeSn films, which would, in turn, lead to improved crystal quality.

The result of an intensive study to find the optimal growth temperature of lattice-matched, intrinsic SiGeSn is shown in the graph in **Figure 14**. For this study, the growth temperature was measured and controlled using a MIR pyrometer.

As it can be seen in the XRD spectrum on the left side, the growth of lattice-matched SiGeSn turns to polycrystallinity at too low growth temperatures ($T_S = 160°C$) and too high Sn concentrations ($c_{Sn} \geq 12.5\%$) due to the reduced surface mobility of the adatoms (see Section 2.1).

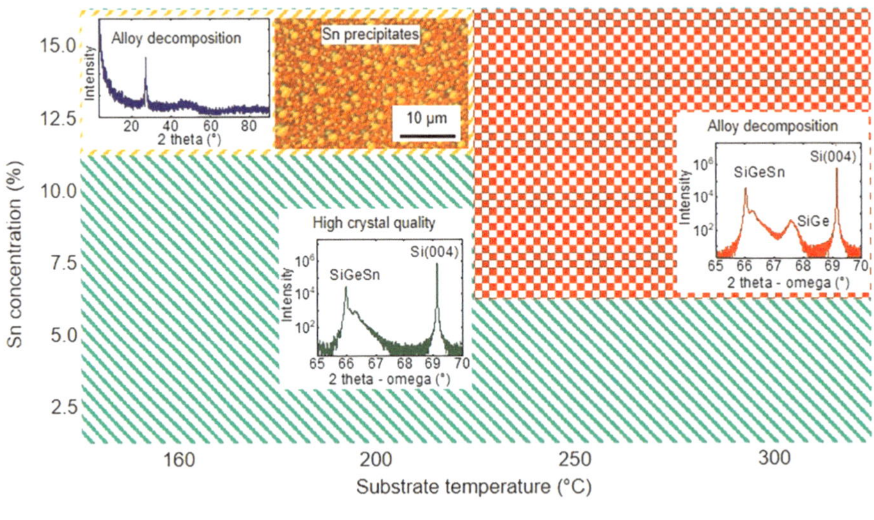

Figure 14.
Overview of the optimal growth temperature of lattice-matched SiGeSn. Green: Regions with satisfactory crystal quality, red: Regions which showed epitaxial breakdown. Orange: Regions with indications for polycrystallinity.

In contrast to that, too high growth temperatures ($T_S \geq 250°C$) not only increases the surface mobility of the adatoms but also strongly intensifies the Sn segregation. Consequently, the incorporation of Sn is reduced so much that the epitaxy process is disturbed, which results in alloy decomposition. Therefore, the XRD spectrum on the right side shows not only a reflection of the SiGeSn film but also of a SiGe film, which does not contain any Sn. In this context, it is often spoken about as an epitaxial breakdown.

The compromise between these extremes can be found at growth temperatures in the range of $T_S = 200°C$ and moderate Sn concentrations ($c_{Sn} \leq 10\%$). Here, the surface mobility of the adatoms is high enough to enable monocrystalline growth. At the same time, the Sn segregation is well limited, so that the epitaxy process is neither disturbed nor broken down. However, if the Sn concentration exceeds $c_{Sn} > 10\%$, the amount of impinging Sn atoms is so high that Sn precipitates can form even at a growth temperature of $T_S = 200°C$. Therefore, other approaches are necessary to achieve a good crystallinity of these Sn-rich SiGeSn compounds. A possible solution would be a reduction of the total growth rate, which would in turn reduce the amount of impinging Sn atoms. Furthermore, the precise control of Sn precipitation is being considered [33].

In contrast to that, low Sn compounds can be grown even at high growth temperatures ($T_S \geq 250°C$) with a satisfactory quality. Therefore, it can be concluded that the absolute amount of impinging Sn atoms seems to have a substantial influence on the resulting crystal quality at a given growth temperature.

4.3 MBE of SiGeSn device structures for electronics and photonics

The previous section reported the optimal growth temperature as the most important growth parameter, for lattice-matched, intrinsic SiGeSn. However, in most cases, not only intrinsic semiconductor regions have to be grown. In order to enable an electronic device functionality, intrinsic regions often alternate with p- or n-type doped regions. Since such regions have substantially different electrical and optical properties, the emissivity ε of the total layer stack, an important parameter for pyrometry, changes drastically with each material transition. Therefore, the measurement signal that changes in this way can no longer be used for control without any doubt because the actual growth temperature is no longer represented properly.

In order to prove this hypothesis, the growth process of a Ge pin diode serves as an example to investigate. Here, the growth temperature was once measured with a thermocouple and also with the MIR pyrometer. The resulting growth temperature characteristics are shown in **Figure 15**.

As it can be seen, the growth temperature, according to the thermocouple signal, remains constant at $T_S = 330°C$ throughout the growth of the p-type doped Ge bottom layer and the intrinsic Ge region. In order to reduce the segregation of Sb (see Section 2.4) during the growth of the n-type doped Ge top layer, the growth temperature was lowered to $T_S = 250°C$. Here, the signal of the thermocouple follows perfectly the desired value. However, at the beginning of the intrinsic region, thus at the interface between p-type doped and intrinsic Ge, the signal of the MIR pyrometer starts to increase strongly. Although the signal then follows the drop of the setpoint, the characteristics show a completely different behavior in the n-type doped region. This can only be explained by a drastic change in the emissivity ε due to each newly introduced interface in the epitaxy layer stack. Consequently, other measurement methods have to be introduced for the growth of device structures, which do not

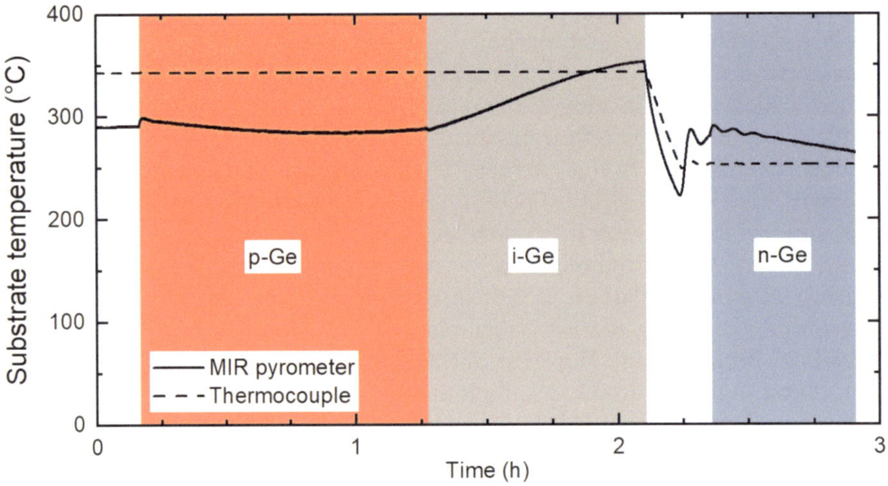

Figure 15.
Growth temperature characteristics of a Ge pin diode measured with a thermocouple (dashed) as well as with a MIR pyrometer (solid).

depend that much on the optical properties of the epitaxial layer stack. Another solution for this problem would be an *in situ* measurement of the emissivity ε, which is, in contrast, quite difficult to realize in the MIR wavelength range of $8\ \mu m \leq \lambda \leq 14\ \mu m$.

However, the MIR pyrometer is, as explained in Section 3.3, still the most suitable method for the measurement of the actual substrate surface temperature at ultra-low growth temperatures.

5. Conclusion

This chapter has presented the fundamentals to understand the physics behind the MBE of group IV compounds. For this, important effects like adsorption, desorption, and the formation of a nucleus have been discussed in detail. Since the epitaxy of group IV compounds automatically includes heteroepitaxy, its basic growth modes, the requirements of monocrystalline growth, and other principles and effects of heteroepitaxy were presented.

Furthermore, this chapter gives a detailed insight into the main components of a group IV compound MBE system, including selection rules for the right molecular beam source suitable for the material to evaporate. Besides that, essential methods for *in situ* monitoring of the growth temperature as well as the optical properties of the epitaxial layers, specifically reflectometry and MIR pyrometry have been presented.

Considering the substantial importance of SiGeSn, especially in the current group IV research, the last part of the chapter covers several technical and physical problems of its MBE. This includes the strongly temperature-dependent segregation of Sn, thus the measurement of the actual surface temperature to suppress it, and the control of the layer composition as well as its homogeneity. In conclusion, MBE is presented as a powerful tool to realize heterostructures and devices containing them based on group IV compounds.

Without a doubt, Ge and SiGeSn have shown to be serious candidates for the wavelength extension of Si photonics from the visual spectrum to the near- and mid-infrared spectrum [31]. Nevertheless, breakthroughs are expected to close the gap with the excellence in material and device performance of group III/V compounds and continuous improvements are necessary to utilize the integration potential of a silicon-based structure. In the following part, we mention some of the activities that promise to have a remarkable input on the ongoing progress.

- Increase of process window for MBE of higher Sn content. Now, the growth temperature is limited to about 200°C, and the postgrowth anneal to below 550°C. Partial ionization of the molecular beam and short-time laser anneal could introduce nonthermal energy contributions that improve ordering without destruction of the single crystallinity. Pavels Onufrijevs [32] reported a direct-indirect band structure formation by Nd:YAG pulsed laser anneal of a low Sn concentration GeSn (4% Sn).

- Stabilizing of the single crystalline diamond lattice by low-dimensional structures (quantum well, quantum wire, quantum dot). Bulk GeSn is only stable up to 1% Sn, but crystalline interfaces stabilize thin epitaxial structures more by up- and down interfaces (wells [35]), all-around interfaces (wires [36]), and island formation (dots).

- Local strain platforms for high strain values. Tensile strain and high Sn content favor the formation of a direct semiconductor GeSn by lowering the energy band gap of the direct conduction sub-band below the indirect bandgap that dominates Ge. The global strain platform (virtual substrate, VS) is not as easily applicable as in the SiGe system because of the limitations in Sn content, whereas in SiGe, the Ge content can be varied up to 100% Ge. Laser concepts in GeSn semiconductors rely mainly on direct bandgap material often supported by local strain produced with micromechanical methods. However, laser functions were demonstrated [37], often with optical drive, low power output, and cooled operation. Although the reasons for the low performance are under discussion, the authors of this chapter believe that the high carrier recombination rate is key to understanding, as explained in the following.

- Reduction of the electronic defect states, which are caused by vacancy supersaturation and residual threading dislocations. Vacancies dominate the point defect equilibrium in Ge and GeSn, whereas interstitials dominate in Si. Misfit dislocations in lattice-mismatched heterostructures are connected to the surface by threading dislocations. High-temperature annealing of the virtual substrate (about 800–850°C) reduces the density of threading dislocations to several 10^7 cm^{-2}. Any further reduction needs special concentration profiles and superlattice filters. The electronic effects of vacancies and threading dislocations were investigated with the Hall effect, with capacitance-voltage (C-V) and deep-level transient spectroscopy (DLTS). Unintentional p-type doping [38] of several 10^{15} cm^{-3}, a mid-gap trap level, and a near valence band level were found. High carrier recombination from band gap levels causes high nonideal dark currents [39, 40], reducing the power of lasers and the detectivity of detectors.

- Internal signal magnification for single photon detection and integration of detector arrays. A proof of concept for single photon detection with a 4×4 Ge array on Si based on SPAD (single photon avalanche diode) is encouraging for infrared night vision [41]. Detector concepts based on phototransistors [42] are integration friendly for large-area arrays. The phototransistors may be based on a hetero-bipolar transistor [43] or on a field effect transistor [43].

Acknowledgements

The authors want to thank all current and former colleagues of the Institute of Semiconductor Engineering of the University of Stuttgart.

Author details

Daniel Schwarz*, Michael Oehme and Erich Kasper
Institute of Semiconductor Engineering, University of Stuttgart, Stuttgart, Germany

*Address all correspondence to: daniel.schwarz@iht.uni-stuttgart.de

References

[1] Wirths S, Buca D, Ikonic Z, Harrison P, Tiedemann AT, Holländer B, et al. SiGeSn growth studies using reduced pressure chemical vapor deposition towards optoelectronic applications. Thin Solid Films. 2014;**557**: 183-187. DOI: 10.1016/j.tsf.2013.10.078

[2] Stange D, von den Driesch N, Rainko D, Roesgaard S, Povstugar I, Hartmann J-M, et al. Short-wave infrared LEDs from GeSn/SiGeSn multiple quantum wells. Optica. 2017;**4**: 185. DOI: 10.1364/OPTICA.4.000185

[3] von den Driesch N, Stange D, Rainko D, Povstugar I, Zaumseil P, Capellini G, et al. Advanced GeSn/ SiGeSn group IV heterostructure lasers. Advanced Science. 2018;**5**:1700955. DOI: 10.1002/advs.201700955

[4] Schulze J, Blech A, Datta A, Fischer IA, Hähnel D, Naasz S, et al. Vertical Ge and GeSn heterojunction gate-all-around tunneling field effect transistors. Solid-State Electronics. 2015; **110**:59-64. DOI: 10.1016/j. sse.2015.01.013

[5] Haehnel D, Fischer IA, Hornung A, Koellner A-C, Schulze J. Tuning the Ge (Sn) tunneling FET: Influence of drain doping, short channel, and Sn content. IEEE Transactions on Electron Devices. 2015;**62**:36-43. DOI: 10.1109/ TED.2014.2371065

[6] Liu M, Schlykow V, Hartmann J-M, Knoch J, Grutzmacher D, Buca D, et al. Vertical heterojunction Ge0.92 Sn0.08/ Ge GAA nanowire pMOSFETs: Low SS of 67 mV/dec, small DIBL of 24 mV/V and highest gm,ext of 870 μS/μm. In: 2020 IEEE Symposium on VLSI Technology. Honolulu, HI, USA: IEEE; 2020. pp. 1-2. DOI: 10.1109/ VLSITechnology18217.2020.9265090

[7] König U, Kibbel H, Kasper E. Si–MBE: Growth and Sb doping. Journal of Vacuum Science and Technology. 1979; **16**:985-989. DOI: 10.1116/1.570165

[8] Ota Y. Silicon molecular beam epitaxy. Thin Solid Films. 1983;**106**:1-136. DOI: 10.1016/0040-6090(83)90180-3

[9] Barnett SA, Greene JE. Si molecular beam epitaxy: A model for temperature dependent incorporation probabilities and depth distributions of dopants exhibiting strong surface segregation. Surface Science. 1985;**151**:67-90. DOI: 10.1016/0039-6028(85)90455-8

[10] Kasper E, Bean JC. Silicon-Molecular Beam Epitaxy. Boca Raton, Fla: CRC Press; 1988

[11] Yamaha T, Nakatsuka O, Takeuchi S, Takeuchi W, Taoka N, Araki K, et al. Growth and characterization of heteroepitaxial layers of GeSiSn ternary alloy. ECS Transactions. 2013;**50**: 907-913. DOI: 10.1149/05009.0907ecst

[12] Asano T, Terashima T, Yamaha T, Kurosawa M, Takeuchi W, Taoka N, et al. Epitaxial growth and crystalline properties of Ge1−−Si Sn on Ge(0 0 1) substrates. Solid-State Electronics. 2015; **110**:49-53. DOI: 10.1016/j.sse.2015. 01.006

[13] Kasper E, Werner J, Oehme M, Escoubas S, Burle N, Schulze J. Growth of silicon based germanium tin alloys. Thin Solid Films. 2012;**520**:3195-3200. DOI: 10.1016/j.tsf.2011.10.114

[14] Oehme M, Kostecki K, Schmid M, Oliveira F, Kasper E, Schulze J. Epitaxial growth of strained and unstrained GeSn alloys up to 25% Sn. Thin Solid Films. 2014;**557**:169-172. DOI: 10.1016/ j.tsf.2013.10.064

[15] Lin H, Chen R, Lu W, Huo Y, Kamins TI, Harris JS. Investigation of the direct band gaps in Ge $_{1-x}$ Sn $_x$ alloys with strain control by photoreflectance spectroscopy. Applied Physics Letters. 2012;**100**:102109. DOI: 10.1063/1.3692735

[16] Li H, Chang C, Chen TP, Cheng HH, Shi ZW, Chen H. Characteristics of Sn segregation in Ge/GeSn heterostructures. Applied Physics Letters. 2014;**105**:151906. DOI: 10.1063/1.4898583

[17] Schwarz D, Funk HS, Oehme M, Schulze J. Alloy stability of Ge1−xSnx with Sn concentrations up to 17% utilizing low-temperature molecular beam epitaxy. Journal of Electronic Materials. 2020;**49**:5154-5160. DOI: 10.1007/s11664-020-08188-6

[18] Olesinski RW, Abbaschian GJ. The Si−Sn (silicon−tin) system. Bulletin of Alloy Phase Diagrams. 1984;**5**:273-276. DOI: 10.1007/BF02868552

[19] Olesinski RW, Abbaschian GJ. The Ge−Sn (germanium−tin) system. Bulletin of Alloy Phase Diagrams. 1984;**5**: 265-271. DOI: 10.1007/BF02868550

[20] Predel B. Ac-Ag ... Au-Zr. Vol. 12A. Berlin/Heidelberg: Springer-Verlag; 2006. DOI: 10.1007/b79358

[21] Kasper E. Growth kinetics of Si-molecular beam epitaxy. Applied Physics A: Materials Science & Processing. 1982;**28**:129-135. DOI: 10.1007/BF00617144

[22] Burton WK, Cabrera N, Frank FC. The growth of crystals and the equilibrium structure of their surfaces. Philosophical Transactions of the Royal Society A London. 1951;**243**:299-358. DOI: 10.1098/rsta.1951.0006

[23] Sze SM, Ng KK. Physics of Semiconductor Devices. 3rd ed. Hoboken, N.J: Wiley-Interscience; 2007

[24] Brownlee LD. Lattice constant of grey tin. Nature. 1950;**166**:482-482. DOI: 10.1038/166482a0

[25] Bauer M, Oehme M, Kasper E. Crystalline to amorphous phase transition in very low temperature molecular beam epitaxy. Materials Science and Engineering: B. 2002;**89**:263-268. DOI: 10.1016/S0921-5107(01)00777-2

[26] Jorke H. Surface segregation of Sb on Si(100) during molecular beam epitaxy growth. Surface Science. 1988;**193**: 569-578. DOI: 10.1016/0039-6028(88)90454-2

[27] Nakagawa K, Sugii N, Yamaguchi S, Miyao M. Ge concentration dependence of Sb surface segregation during SiGe MBE. Journal of Crystal Growth. 1999;**201–202**:560-563. DOI: 10.1016/S0022-0248(98)01389-X

[28] Sawano K, Hoshi Y, Kasahara K, Yamane K, Hamaya K, Miyao M, et al. Ultrashallow ohmic contacts for n-type Ge by Sb δ-doping. Applied Physics Letters. 2010;**97**:162108. DOI: 10.1063/1.3503587

[29] Jorke H, Herzog H-J, Kibbel H. Secondary implantation of Sb into Si molecular beam epitaxy layers. Applied Physics Letters. 1985;**47**:511-513. DOI: 10.1063/1.96109

[30] Jorke H, Kibbel H. Doping by secondary implantation. Journal of the Electrochemical Society. 1986;**133**: 774-778. DOI: 10.1149/1.2108674

[31] Kasper E, Yu J. Silicon-Based Photonics. 1st ed. New York: Jenny Stanford Publishing; 2020. DOI: 10.1201/9781315156514

[32] Onufrijevs P, Ščajev P, Medvids A, Andrulevicius M, Nargelas S, Malinauskas T, et al. Direct-indirect GeSn band structure formation by laser radiation: The enhancement of Sn solubility in Ge. Optics & Laser Technology. 2020;**128**:106200. DOI: 10.1016/j.optlastec.2020.106200

[33] Takeuchi S, Shimura Y, Nakatsuka O, Zaima S, Ogawa M, Sakai A. Growth of highly strain-relaxed Ge1−xSnx/virtual Ge by a Sn precipitation controlled compositionally step-graded method. Applied Physics Letters. 2008;**92**:231916. DOI: 10.1063/1.2945629

[34] Margrave J. The Characterization of High-Temperature Vapors. New York: John Wiley & Sons, Incorporated; 1967

[35] Oehme M, Kasper E, Weißhaupt D, Sigle E, Hersperger T, Wanitzek M, et al. Two-dimensional hole gases in SiGeSn alloys. Semiconductor Science and Technology. 2022;**37**:055009. DOI: 10.1088/1361-6641/ac61fe

[36] Kim Y, Assali S, Joo H-J, Koelling S, Chen M, Luo L, et al. Short-wave infrared cavity resonances in a single GeSn nanowire. Nature Communications. 2023;**14**:4393. DOI: 10.1038/s41467-023-40140-0

[37] Zhou Y, Ojo S, Wu C-W, Miao Y, Tran H, Grant JM, et al. Electrically injected GeSn lasers with peak wavelength up to 2.7 μm. Photonics Research. 2022;**10**:222. DOI: 10.1364/PRJ.443144

[38] Tetzner H, Seifert W, Skibitzki O, Yamamoto Y, Lisker M, Mirza MM, et al. Unintentional p-type conductivity in intrinsic Ge-rich SiGe/Ge heterostructures grown on Si(001). Applied Physics Letters. 2023;**122**:243503. DOI: 10.1063/5.0152962

[39] Ghosh S, Sun G, Morgan TA, Forcherio GT, Cheng H-H, Chang G-E. Dark current analysis on GeSn p-i-n photodetectors. Sensors. 2023;**23**:7531. DOI: 10.3390/s23177531

[40] Chang G-E, Yu S-Q, Sun G. "GeSn Rule-23"—The performance limit of GeSn infrared photodiodes. Sensors. 2023;**23**:7386. DOI: 10.3390/s23177386

[41] Wanitzek M, Oehme M, Schwarz D, Guguieva K, Schulze J. Ge-on-Si avalanche photodiodes for LIDAR applications. In: 2020 43rd International Convention on Information, Communication and Electronic Technology (MIPRO). Opatija, Croatia: IEEE; 2020. pp. 8-12. DOI: 10.23919/MIPRO48935.2020.9245425

[42] Kumar H, Basu R. Study of the effect of temperature on the detectivity and sensitivity of GeSn-based heterojunction phototransistor for mid-wave infrared applications. Applied Physics B: Lasers and Optics. 2021;**127**:13. DOI: 10.1007/s00340-020-07569-3

[43] Agarwal B, Kumar H. Novel group-IV alloy-based MOS field-effect phototransistors for near-infrared applications. IEEE Sensors Journal. 2023;**23**:16797-16804. DOI: 10.1109/JSEN.2023.3287880

Section 2
Magnetic Characterization

Chapter 5

Unraveling the Extraordinary Anisotropic Magnetoresistance in Antiferromagnetic Perovskite Heterostructures: A Case Study of $CaMnO_3/CaIrO_3$ Superlattice

Suman Sardar

Abstract

Antiferromagnetic (AFM) spintronics offers advantages over ferromagnetic (FM) spintronics, such as zero stray fields, closer packing, and imperviousness to disruptive fields. Anisotropic magnetoresistance (AMR) can be enhanced by materials with pronounced spin-orbit coupling (SOC) and magnetocrystalline anisotropies. AMR research aims to develop new materials and heterostructures with enhanced and tunable anisotropic transport properties for advanced electronic devices. The nonmagnetic ground state of iridium pseudospin moments in $SrIrO_3$ and $CaIrO_3$ is determined by SOC and electron correlations (U). This study shows that by coupling $CaIrO_3$ with a severely distorted canted AFM manganite $CaMnO_3$, the AMR can be increased by more than one order of magnitude, primarily due to interlayer coupling. Additionally, the spin-flop transition in a nearly Mott region contributes to an unprecedented AMR of 70%, two orders of magnitude larger than previously achieved. The study demonstrates that thin films of canted AFM phases of $CaMnO_3$ and $CaIrO_3$ exhibit dimensionality control, with a diminishing magnetic moment, and the valence state can be altered at interfaces in superlattices involving manganites.

Keywords: transition metal oxides and heterostructures, antiferromagnetic spintronics, anisotropic magnetoresistance, spin-orbit coupling, canted antiferromagnet, spin-flop transition

1. Introduction

Antiferromagnetic (AFM) spintronics has emerged as a promising alternative to ferromagnetic (FM) spintronics due to several advantages. AFM spintronics offers a significant advantage over other technologies as its ordered microscopic moments alternate between individual atomic sites, resulting in zero stray fields and enabling closer packing of devices without interference. Thus, AFM materials possess a remarkable property of being externally invisible in magnetic fields, owing to their zero net magnetic moments, which implies that any information stored in their magnetic moments would be impervious to disruptive magnetic fields and beyond the detection capabilities of commonly used magnetic probes [1–6]. Antiferromagnets have been discovered to have greater potential than just being used as passive elements in exchange bias applications. When a soft ferromagnet is placed next to a thin antiferromagnet, and a feeble external magnetic field is applied, the ferromagnet can reposition itself, causing its interfacial moments to pull the adjacent antiferromagnetic moments via the interfacial exchange spring. Metal antiferromagnets, such as the Ir-Mn alloys, are versatile materials with various applications, ranging from serving as advantageous passive exchange-bias components to functioning as active electrodes in Tunneling Anisotropic Magnetoresistance (TAMR) devices [7–11]. Additionally, AFM spintronics devices require lower operational power, which makes them more energy-efficient. AFM spintronics allows for ultrafast control of the staggered spins at terahertz frequencies, which can enable faster and more efficient data processing. Current-induced spin-orbit torque (SOT) and anisotropic magnetoresistance (AMR) are two important phenomena used in AFM spintronics devices and in this way, the spin of the AFM material can be accessed for data writing and reading operations. SOT is a process that utilizes electric current to generate torque on the magnetization of a material, primarily used in AFM spintronic devices. The electron's spin polarization generates a spin current when a current is passed through a thin layer of heavy metal. The interaction of the spin current with the AFM moments at the interface generates a torque that can reorient the magnetic moments of the AFM layer, enabling precise data writing and reading operations by controlling the current's direction and magnitude. Moreover, the SOT-based devices have the advantage of low power consumption and fast switching speeds making them promising candidates for future high-density data storage and processing applications [6, 10, 12]. AMR is a phenomenon that causes the resistance of a material to change as a function of the angle between the direction of current flow and the magnetization direction. Overall, AFM spintronics has the potential to revolutionize the field of spintronics and enable the development of faster, more efficient, and more compact devices [13–15].

Here, a detailed discussion of AMR will be given in 3d/5d perovskite heterostructures. The origin of AMR depends on whether the material is crystalline or noncrystalline. In both cases, the AMR response arises from the anisotropy of the electron scattering or transport properties. In crystalline materials, AMR arises from the anisotropy of the electron scattering rate, which is influenced by the orientation of the magnetic field with respect to the crystal lattice [14, 16, 17]. As an example, a high electronic band coupling to crystal sites can lead to a stronger AMR response and itinerant electrons from higher energy bands typically contribute to the magnetic effects in AMR, resulting in a greater change in electrical resistance in $LaAlO_3/SrTiO_3$ (111) heterostructure [18]. On the other hand, in noncrystalline materials such as amorphous alloys, the AMR originates from the anisotropy of the electron transport properties. Researchers working on AMR seek to design heterostructures with specific properties that can enhance the AMR effect. One approach is to use materials with pronounced SOC and magnetocrystalline anisotropies. This can cause the anisotropies to manifest as anisotropic transport. In some cases, the magnetic field and weak magnetic moments of the canted AFM phase can couple together, resulting in emergent magnetic and topological properties in oxide heterostructures. For example, 3d-5d oxide heterostructures such as epitaxial superlattices (SLs) of iridium oxides $(Ca/SrIrO_3)/SrTiO_3$ and manganite/iridate SLs, have been found to exhibit large atomic SOC and electron-correlation-dominated AMR [19]. Furthermore, current-dependent AMR and MR offer an opportunity to explore momentum-dependent scattering to elucidate the role of Rashba SOC. The Rashba SOC introduces spin splitting along the momentum axis, whereas atomic SOC does not [20]. However, a few research groups have proposed that besides the atomic SOC, the Rashba SOC also plays an important role in the AMR in transport. The primary focus of AMR research is to design and develop new materials and heterostructures that can exhibit enhanced and tunable anisotropic transport properties for use in advanced electronic devices.

The nonmagnetic ground state of iridium (Ir) pseudospin moments in $SrIrO_3$ and $CaIrO_3$ is determined by the interplay between SOC and electron correlations (U). The value of U, which represents the strength of the Coulomb interaction between electrons, affects the electronic and magnetic properties of the system. When U is large, the system tends to favor a Mott insulating state, which is characterized by the localization of electrons due to strong Coulomb repulsion. On the other hand, when U is small, the system can exhibit metallic behavior. In epitaxial thin films, the dimensionality of the system can be tuned by controlling the thickness of the film. It has been found that reducing the dimensionality in epitaxial thin films can increase U and induce pseudospin-based emergent magnetism in $SrIrO_3$ and $CaIrO_3$ [21–24]. This emergent magnetism arises from the system's delicate balance between SOC and U. In a low-bandwidth SL, i.e., $CaIrO_3/SrTiO_3$, the AMR effect is attributed to a combination of different factors including in-plane biaxial magnetic anisotropy, magneto-elastic coupling, and interlayer exchange coupling based on tilted oxygen octahedra with glazer notation $(a^-a^-c^+)$ across the constituent layers [19, 25, 26]. Despite the concerted efforts to enhance the AMR effect in 3d-5d heterostructures, the maximum amplitude of the fourfold AMR signal is still limited to 1%. This limitation highlights the need for the development of new strategies for enhancing the AMR effect, such as the careful selection of constituent materials and the optimization of heterostructure architecture to get a larger AMR effect. This study shows that by coupling the $CaIrO_3$ with a severely distorted canted AFM manganite $CaMnO_3$ and using the same sense of $(a^-a^-c^+)$ oxygen octahedra tilts, the AMR can be increased by more than one order of magnitude which is about 20%. This increase is mainly generated due to the interlayer coupling of the $CaMnO_3/CaIrO_3$ layer. Additionally, the spin-flop transition in a nearly Mott region triggers an additional two-order of AMR amplitude along with a four-fold symmetry component. By combining these two effects, an exceptional AMR of 70% has been achieved, surpassing the previous record by two orders of magnitude. The study also shows controls and other unique facets related to these effects. The study demonstrates that thin films of canted AFM phases of $CaMnO_3$ and $CaIrO_3$ exhibit dimensionality control, with a diminishing magnetic moment. The M-T curve shows a signature of a magnetic transition occurring around 70–100 K, representing an AFM phase transition in the $CaMnO_3/CaIrO_3$ SLs. The valence state can be altered at interfaces in SLs involving manganites, resulting in emergent magnetic phenomena through the transfer of charge. Research conducted on $CaRuO_3/CaMnO_3$ SLs has revealed that electron leakage into the $CaMnO_3$ layer decreases exponentially from the interface to the bulk of the layer [27]. From the exchange bias data of $CaMnO_3/CaIrO_3$ SLs, it is clear that a charge transfer from the $CaIrO_3$ to $CaMnO_3$ layer near the interface develops FM and AFM phases at the interface and bulk of $CaMnO_3$ layers, respectively.

2. Experimental section

2.1 Sample synthesis and structural characterizations

Pulsed laser deposition (PLD) is a valuable experimental technique in condensed matter physics for investigating interface properties [27, 28]. Creating a well-defined interface and high-quality thin film is essential, and PLD involves laser ablation of target material, plasma plume dynamics, and nucleation and growth of atoms on the substrate surface. Laser energy selection is critical based on the ablation threshold energy of bulk materials, and a KrF excimer laser with a wavelength of 240 nm is commonly used in lab experiments. The gas pressure is used to control plume dynamics and reduce its energy, resulting in a dense and unidirectional plume. The nucleation of atoms on the substrate is a crucial step in crystal formation, and the suitable combination of laser energy, gas pressure, and substrate temperature determines the mobility of surface atoms. Oxygen gas is supplied inside the PLD chamber from an external cylinder via an inlet to produce an oxide film, and different growth mechanisms, depending on the mobility of surface atoms and surface smoothness, occur on the surface (**Figure 1**).

Reflection high energy electron diffraction (RHEED) technology is used to study the growth mechanism, providing valuable information about the surface morphology and structure. RHEED enables the monitoring of various types of growth such as layer-by-layer (2D) growth and island (3D) growth. The process involves nucleation and growth of atoms on the substrate surface which can be visualized using the RHEED set-up. High-pressure RHEED has been developed to monitor surface structure during oxide deposition at higher pressures, opening up new possibilities for material synthesis. To obtain the atomically sharp TiO_2-terminated surface which is essential for layer-by-layer growth, the substrates were treated in deionized water and buffered NH_4F-HF (BHF) solution [30], followed by thermal annealing at 960°C for 1.5 hours. The SLs were grown on a substrate maintained at a temperature of 730°C and under an oxygen partial pressure of 6 Pa. Subsequently, the samples underwent a post-annealing process at identical temperature and pressure conditions for a duration of 30 minutes. The thickness of the SLs was precisely controlled and monitored by using in-situ RHEED. The high structural quality of the SLs was examined in detail by

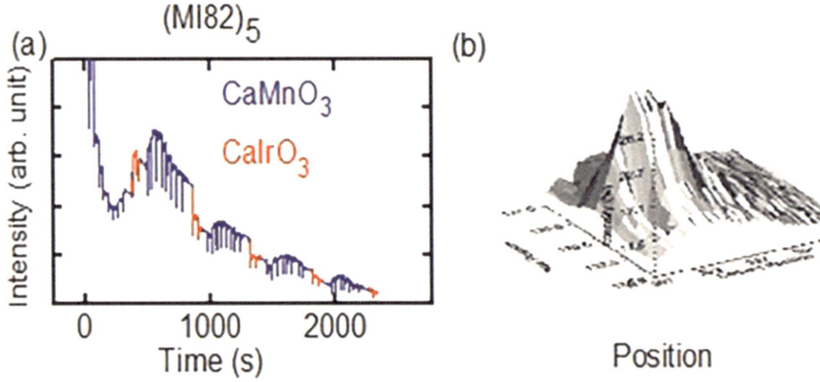

Figure 1.
Pulsed laser interval deposition [29]. (a) How the oscillation intensity changes with the increasing number of laser shots and the formation of each layer represented by one complete oscillation. (b) RHEED specular spot intensity and its surface plot in X-Y coordinates.

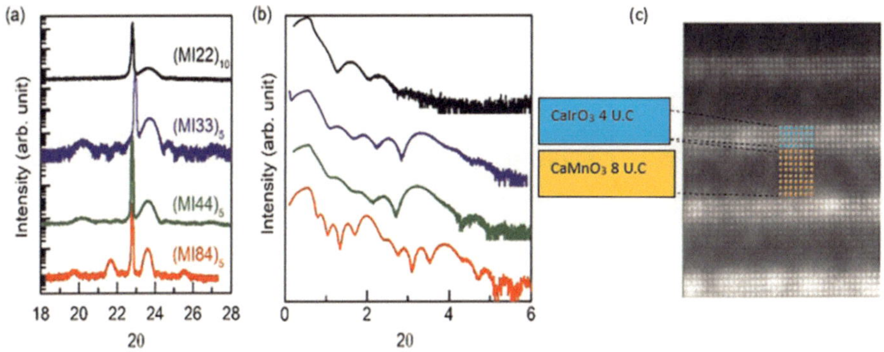

Figure 2.
(a, b) Scan of θ-2θ and X-ray reflectivity for (MIxy)z with x = y = 2–4, $(MI22)_{10}$, $(MI33)_5$, $(MI44)_5$, and $(MI84)_5$ SLs. (c) Cross-sectional HAADF-STEM image that exhibits the atomic level resolution of $CaMnO_3$ and $CaIrO_3$ layers in the $(MI84)_5$ SLs is shown. (Reprinted from [14] © 2023 American Physical Society).

performing room temperature X-ray diffraction (XRD) scans using a PANalytical X'pert Pro diffractometer. Finally, the thickness of the SLs was confirmed by using X-ray reflectivity (XRR).

The samples are $CaMnO_3/CaIrO_3$ SLs with different configurations based on the periods of constituent layers. The SLs are coded as (MIxy)z, where M and I refer to $CaMnO_3$ and $CaIrO_3$ layers, respectively. The value of 'x' and 'y' represents the period or unit cell (u.c.) of the $CaMnO_3$ and $CaIrO_3$ layers, respectively, and 'z' represents the number of repetitions. The article discusses the structural properties and variations in AMR of $CaMnO_3/CaIrO_3$ SLs. The samples are categorized based on the period of constituent layers as (I) The first category involves the simultaneous variation of $CaMnO_3$ and $CaIrO_3$ u.c., with samples including (MIxy)z for x = y = 2–4, and $(MI84)_5$ with a larger $CaMnO_3$ period; (II) The second category involves the control of AMR by fixing $CaIrO_3$ period while varying the $CaMnO_3$ period, with samples $(MIx2)_5$ for x = 2,4,6,8 and $(MIx4)_5$ for x = 2,4,6; (III) The third category involves keeping the $CaMnO_3$ period fixed and varying the $CaIrO_3$ period, with samples (MI2y)z for y = 2,4,5 and (MI5y)z for y = 2,5,8. The SLs were formed by using a pulsed interval deposition technique and analyzed by using XRD and HAADF-STEM. In **Figure 1(a)**, the synthesis of (MI82)5 SL is shown, where one complete oscillation represents the formation of one u.c. Additionally, the Gaussian pattern of the specular spot intensity after finishing the deposition confirms sharp, layer-by-layer growth, as shown in **Figure 1(b)**. **Figure 2 (a)** and **(b)** provide XRD and XRR data, respectively, for the primary series and higher periodic SL, i.e., $(MI84)_5$ SL. Also, **Figure 2 (c)** shows HAADF-STEM data for $(MI84)_5$ SL. These data offer insights into the crystal structure, periodicity, and interface quality of the SLs.

2.2 Investigating magnetic-properties of interface

2.2.1 Magnetization measurements, X-ray absorption study, and understanding charge transfer and spin canting mechanism across $CaMnO_3/CaIrO_3$ Heterointerface

The magnetization (M) versus temperature (T) and magnetic field (H) data of (MIxy)z SLs with x = y = 2–4 are plotted in **Figure 3 (a)** and **(b)**. As the number of periods increases, there is a noticeable decrease in both the magnetic transition

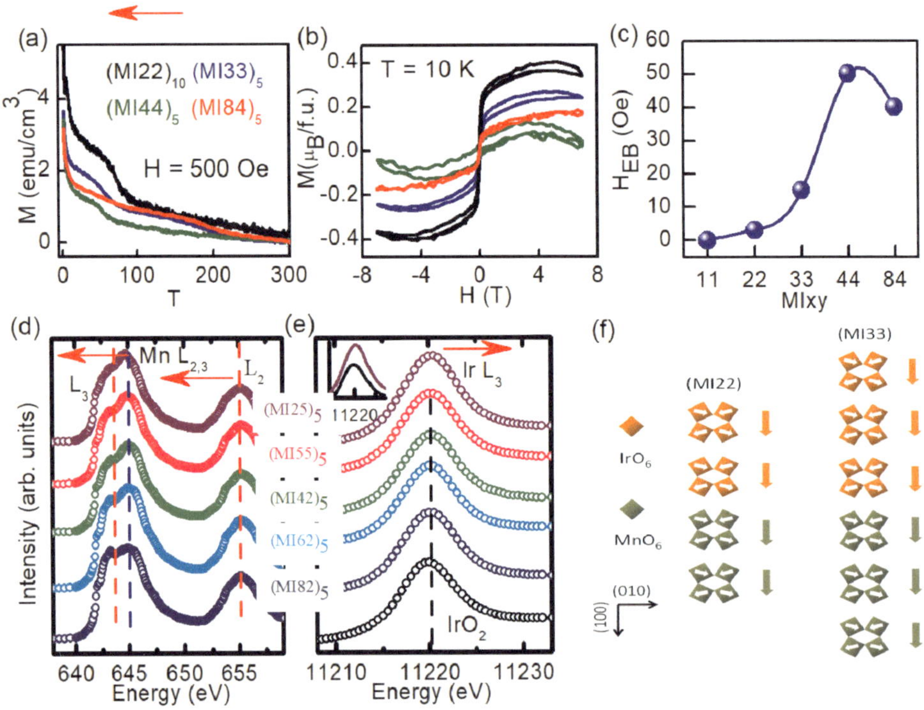

Figure 3.
Displays (a) the temperature (T), (b) field (H) dependence of magnetization (M), and (c) the strength of the exchange bias field (H_{EB}) for (MIxy) SLs with varying stacking of $CaMnO_3$ and $CaIrO_3$ layers. X-ray absorption spectra around (d) the Mn $L_{2,3}$ edge (TEY modes) and (e) Ir L_3 edge (FY modes) are also presented for (MIxy)z SLs, where the insets show a comparison of the $(MI25)_5$ SL with the IrO_2 sample. (f) a schematic illustration is shown, demonstrating the spin canting in the $CaIrO_3$ and $CaMnO_3$ layers. CaIrO3 and CaMnO3 layers are organized into BO6 planes that are arranged in the ab plane and stacked along [001]. To simplify the explanation, a top-down view of the canted moments and net magnetic moment is shown along two in-plane directions, namely [010] and [100], with respect to the pseudocubic STO (100) substrate. The length of the arrows showing the net magnetic moment does not represent the relative size of the magnetic moment in $CaIrO_3$ or $CaMnO_3$ layers. (Reprinted from [14] © 2023 American Physical Society).

temperature (Tc) and the saturation magnetic moment (Msat), as depicted in **Figure 3**. For example, the $(MI22)_{10}$ SL exhibits Tc of approximately 100 K, while the Tc decreases to approximately 60 K for $(MI44)_5$ SL and vanishes for the higher periodic $(MI84)_5$ SL. This trend suggests that the magnetic properties of the SLs are strongly influenced by the periodicity of the constituent layers, with larger periods leading to weaker magnetism. The M-H data in **Figure 3 (b)** shows a saturation magnetic moment (Msat) of approximately 0.4 μB/f.u. for $(MI22)_{10}$ SL, which value is consistent with the reported canted AFM state in $CaIrO_3/CaMnO_3$ heterostructures [31]. To further investigate the magnetic properties of the SLs, the exchange-bias fields (H_{EB}) were measured by performing field-cooled M-H measurements, as shown in **Figure 3(c)**.

The interfacial charge transfer and electronic structure near the Fermi level were examined using x-ray absorption spectroscopy (XAS) at both the Mn and Ir L-edges. The spin-orbit coupled states are obtained in the spectra containing two features, the $L_3(2p_{3/2})$ and $L_2(2p_{1/2})$ edges, respectively, in Mn 2p core hole, as shown in **Figure 3 (d)**. A clear shift of the L_3 edge of the spectra towards lower energy with respect to the other samples was observed in the $(MI25)_5$ SL, indicating the presence of Mn^{3+} ions, which were likely formed by the transfer of charge from the Ir to the Mn at the interface. This shift was evident in both the surface-sensitive total electron yield

(TEY) and bulk-sensitive fluorescence yield (FY) mode for Mn-edge and Ir-edge spectra, respectively, as shown in **Figure 3(d, e)** and **4**. Thinner $CaMnO_3$ layers in SLs receive a larger fraction of electrons, causing a larger shift in the Mn edge for $(MI25)_5$ with 2 u.c. of $CaMnO_3$ period compared to SLs with thicker $CaMnO_3$ layers compared to $CaIrO_3$, as shown in **Figure 3 (d)**. Additionally, the larger thickness/volume of $CaIrO_3$ in $(MI25)_5$ SL will transfer a larger number of electrons to $CaMnO_3$. The observed minimal shift in the Ir edge, as shown in **Figures 3(e)** and **4**, suggests that the Ir^{4+} state is highly stable, and only a small amount of charge transfer occurs across the interface. Therefore, the remaining change in the Mn valency is assumed to arise from vacancies in the manganese layer. Approximately, 0.1 hole/electrons per Ir/Mn ion are transferred at $CaMnO_3/CaIrO_3$ interfaces for the $(MI82)_5$ SL, while the remainder of the change in the Mn valence is due to vacancies in the manganese layer and, also, the charge transfer is sensitive to the constituent layer thickness, with a maximum of approximately 0.25 hole/electron per Ir/Mn ion transferred for the $(MI25)_5$ SL, as shown in the table-**I**. It is observed from the table that there is charge transfer from Ir to Mn assuming a valency range of ~3.8–3.95 for the bare $CaMnO_3$ layer. As the $CaIrO_3$ layer thickness increases and becomes comparable with $CaMnO_3$ layer thickness, such as in MI25, MI44, and MI42 samples, the fraction of charge transfer increases, contributing to the larger conductivity and number of carriers available for charge-transfer in thicker $CaIrO_3$ layers. This is evident in the comparison between MI25 and MI62 samples (**Table 1**).

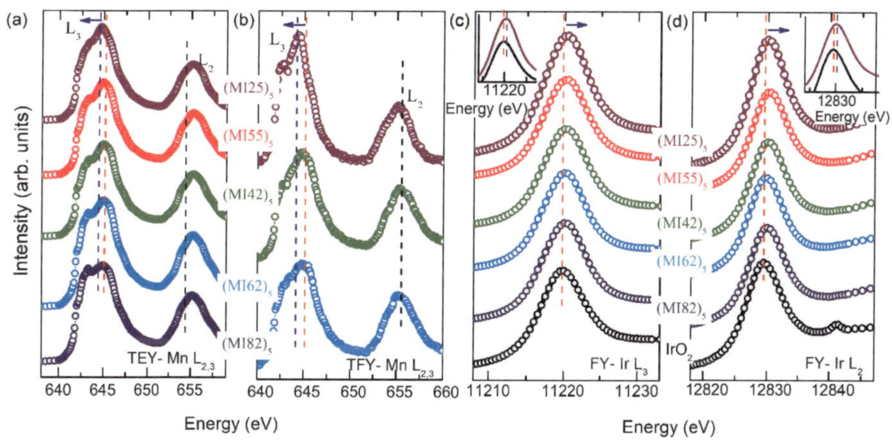

Figure 4.
XAS spectra around the Mn $L_{2,3}$ edges for $[(CaMnO3)_x/(CaIrO3)_y)]_z$ measured in the (a) TEY and (b) TFY modes. XAS spectra of the Ir $L_{2,3}$-edges for [(CMOx/CIOy)]z measured in the fluorescence yield modes. Insets depict the comparison view of $[(CaMnO_3)_2/(CaIrO_3)_5)]_5$ with the reference IrO_2 sample. (Reprinted from [14] © 2023 American Physical Society).

Sample	MI82	MI62	MI42	MI44	MI25
Mn edge valency	3.68(4)	3.78(8)	3.79(2)	3.74(4)	3.59(4)
Ir edge valency	4.09(7)	4.05(7)	4.18(1)	4.17(4)	4.25(9)

Table 1.
Charge transfer across $CaMnO_3/CaIrO_3$ interface.

The charge transfer between $CaIrO_3$ and $CaMnO_3$ layers depends on the volume or the number of available carriers in $CaIrO_3$ and the number of $CaMnO_3$ layers. Due to the vacant 'e_g' orbital near the Fermi level, $CaMnO_3$ with a distorted lattice exhibits a significant attraction towards electrons. In another study on Ce^{4+} – doped $CaMnO_3$, even a small electron doping induced canting of the AFM lattice resulting in a significant increase in magnetic moments [32]. Leakage of electrons into $CaMnO_3$ decay exponentially from the interface to the bulk of the layer in $CaRuO_3/CaMnO_3$ SLs [27, 33]. The interface layer receives the maximum charge density, inducing a double-exchange governed FM phase or a largely canted AFM phase at the $CaMnO_3$ interface layer. The deeper $CaMnO_3$ layer tends to remain AFM. The formation of a magnetic gradient across the $CaMnO_3$ layer is well supported by both theory and experiments [33]. The H_{EB} is the signature of FM and AFM phases across the interface. As depicted in **Figure 3(c)**, the (MI22)10, (MI33)5, (MI44)5, and (MI84)5 SLs display H_{EB} of 3, 15, 50, and 35 Oe, respectively. To achieve a higher H_{EB}, a virtual arrangement of the FM/AFM interface is necessary, which is absent in the first SL since its $CaMnO_3$ layers are only present at the interface. However, in the latter three SLs, the $CaMnO_3$ layer thickness increases along with the FM interface, resulting in the manifestation of H_{EB} which increases with the thickness of the $CaMnO_3$ layer. This result can be used to look for the formation of such magnetic gradients also where the AFM state apart from the interface remains unchanged for both $CaIrO_3$ and $CaMnO_3$.

2.3 Study on anisotropic magnetoresistance (AMR)

2.3.1 ϕ-AMR in (MIxy)z (x = y = 2: 4)

The AMR, also referred to as angular dependent magnetoresistance, was measured in three different senses of rotations of the SLs with respect to the magnetic field, as shown in **Figure 5(a)**, and is calculated as

$$AMR = \frac{\rho[B\,(angle)] - \rho[B\,(angle = 90°)]}{\rho[B\,(angle = 90°)]} \tag{1}$$

Where 'angle' represents the angle between the magnetic field and the current direction.

In **Figure 5 (a)**, H rotates in the xy, yz, and zx planes while three rotation angles, namely ϕ, θ, and γ, are depicted. Additionally, the current direction is along the (100) axis of the sample. **Figure 5 (b-f)** depicts the magnetic and electrical behavior of (MIxy)z (x = y = 2–4), which can be explained by the enhancement of 'U' induced by dimensionality and the charge transfers. All SLs demonstrate diminishing AMR in the range of 70–100 K which is the same as the magnetic transition temperature (Tc) of the SLs. The sheet resistance was also measured, and it was found to increase with decreasing $CaIrO_3$ period, as shown in **Figure 5 (b)**. The Mott state is characterized by a sudden increase in resistivity below 50 K. As the $CaIrO_3$ period decreases, the sheet resistance increases and the resistance of a 10 nm $CaIrO_3$ film is clearly distinct from the other SLs. In particular, $(MI22)_{10}$ tends to exhibit a Mott-type state below 30 K. These SLs' emergent magnetic and transport properties are related to their AMR, which will be discussed further below. Now, it is important to investigate AMR properties of several 3d-5d SLs including $(MI22)_{10}$, $(MI33)_5$, $(MI44)_5$, and $(MI84)_5$ as

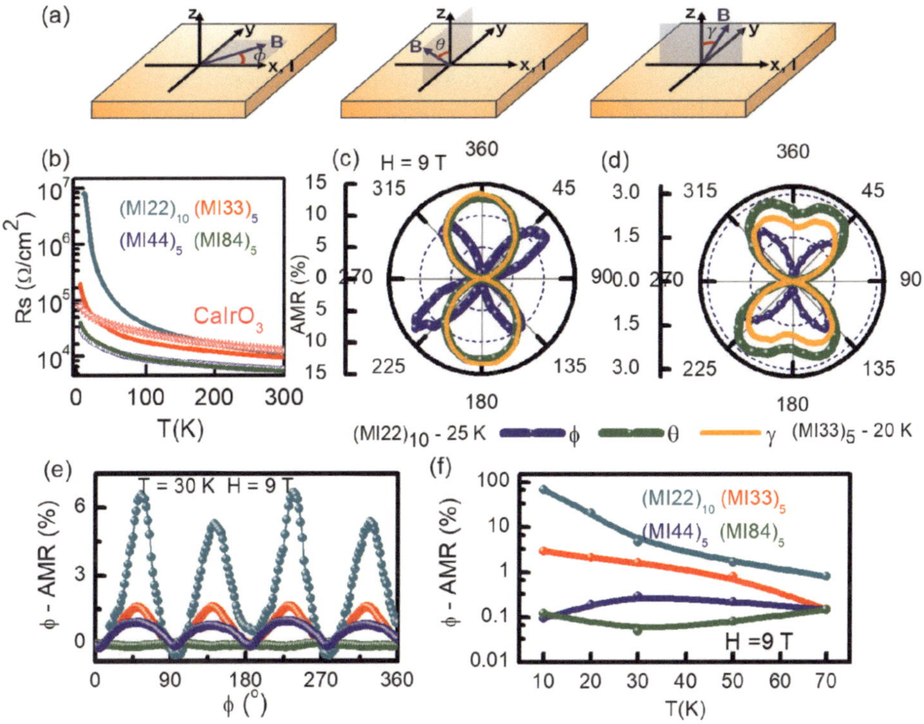

Figure 5.
(a) Three different rotational geometries schematically illustrating the measurements of AMR. (b) The impact of extended $CaMnO_3$ and $CaIrO_3$ periods on the sheet resistance was examined by plotting the temperature-dependent behavior of (MIxy)z (x = y = 2–4) SL and (MI84)5 SL. (c) Polar plots comparing the ϕ, θ, and γ AMRs are presented for $(MI22)_{10}$ and (d) $(MI33)_5$ SL. (e) The variation in ϕ-AMR at 30 K is demonstrated for both (MIxy)z (x = y = 2–4) and $(MI84)_5$ SL. (f) The variation in the ϕ-AMR amplitude as a function of temperature for H = 9 T is shown. (Reprinted from [14] © 2023 American Physical Society).

a function of temperature, magnetic field, and period of SLs. The AMR measurements reveal that the ϕ-AMR exhibits four-fold sinusoidal oscillations for both $(MI22)_{10}$ and $(MI33)_5$, with a subtle two-fold component superimposed on a dominant four-fold component, shown in **Figure 5 (e)**. In **Figure 5 (f)**, the ϕ-AMR in $(MI22)_{10}$ SL is particularly striking, exhibiting an astounding 70% amplitude at 10 K, which decreases to 23% at 20 K and gradually ceases to manifest at 100 K. This is the largest amplitude of four-fold ϕ-AMR reported not only for 3d-5d SLs but also for other oxide heterostructures. Furthermore, the ability to tune the ϕ-AMR by two orders of magnitude is demonstrated by varying the period of the SLs. The θ- and γ-AMRs, shown in **Figure 5 (c, d)**, also show a phenomenal amplitude of 15% for $(MI22)_{10}$, which is much larger than observed for any 3d-5d SL. The AMR decreases with increasing temperature and completely disappears around the transition temperature in the range of 70–100 K for all samples. The ϕ-AMR observed in $(MI22)_{10}$ SL, as shown in **Figure 5 (f)**, is exceptional because it is much larger than what has been reported in any 3d-5d heterostructures so far, and it is the largest among complex oxide heterostructures. Additionally, the ϕ-AMR reduces by one order of magnitude as the period (x = y) increases from 2 to 4. This sensitivity to the constituent layer thickness suggests the presence of a unique phenomenon that promotes interlayer coupling.

2.3.2 The origin of the AMR

One method for determining the origin of AMR is to compare the behavior of different types of AMRs and analyze their dependence with the magnetic field. In particular, the ϕ, θ, and γ AMRs have been studied extensively.

3. Non-quadratic dependence of ϕ-AMR with field

One observation that has been made is that none of the ϕ, θ, and γ AMRs follow a quadratic dependence on the magnetic field because ϕ-AMR does not vary linearly with B^2, as shown in **Figure 6**. These results rule out Lorentz scattering as the origin of the effect. Lorentz scattering is a mechanism in which the magnetic field causes the scattering of electrons, leading to a change in resistance [19]. Another observation that has been made is that the magnitude and phase of the θ and γ AMRs for the sample $(MI22)_{10}$ are coincident. This eliminates the possibility of spin Hall MR or s-d scattering as the underlying mechanisms. Spin Hall MR is a mechanism in which the spin Hall effect causes a change in resistance, while s-d scattering is a mechanism in which the scattering of electrons between the spin-polarized d-band and s-band causes a change in resistance [34–37].

3.1 Intra and interlayer coupling in ϕ-AMR for varying periodicity: (MIx2)z for x = 2, 4,6 and 8 and (MI2y)z for y = 2,4 and 5 SLs

In the canted AFM phase of these SLs, interlayer coupling controls the domain scattering mechanism based on biaxial magnetic anisotropy, which is believed to be the underlying cause of the AMR. Domain scattering refers to the scattering of electrons as they pass through the domains (regions) with different magnetic orientations in a material. In this case, the domains are created by the canted AFM phase of the SLs, and the domain scattering is thought to be responsible for AMR observed in the material. The magnitude of interface coupling and the dimensions of the individual layers are determining factors in the strength of interlayer coupling. The magnetic interactions for intra- and interlayer bonds of neighboring iridium ions can be expressed in a common form [eq. (2)], with different coupling constants for inter- and intralayer bonds [38].

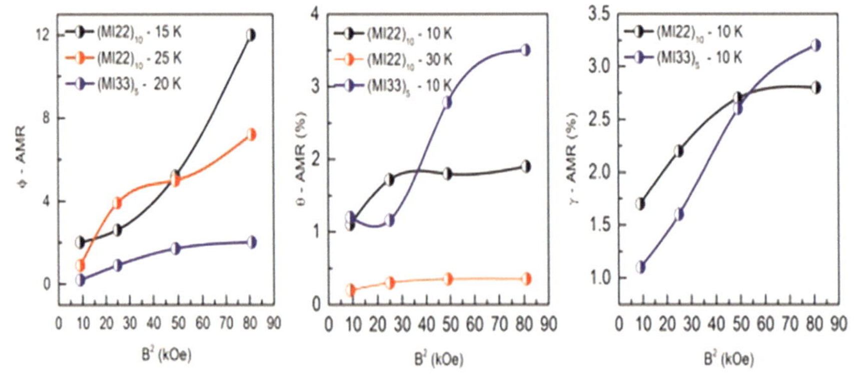

Figure 6.
At various temperatures, the $(MI22)_{10}$ and $(MI33)_5$ SL display non-quadratic dependence of B as seen in the polar plots of ϕ, θ, and γ - AMRs presented. (Reprinted from [14] © 2023 American Physical Society).

$$H_{ij} = J_{ij}\vec{S}_i.\vec{S}_j + J_{1c}\vec{S}_i.\vec{S}_j + \Gamma_{ij}\vec{S}_i^{\,z}.\vec{S}_j^{\,z} + \overrightarrow{D}_{ij}.\left[\vec{S}_i \times \vec{S}_j\right] \quad (2)$$

Where J_{ij} is the in-plane isotropic Heisenberg exchange among pseudospin $\vec{S}_i$ *and* $\vec{S}_j$, and J_{1c} represents the first interlayer interaction. In the case of 5d material, pseudospin is an entangled state of spin and orbital moment denoted as J_{eff} [39]. Γ_{ij} represents symmetric exchange anisotropy that favors collinear c-axis spin order, and $\vec{D}_{ij}$ whose direction is along the c-axis represents antisymmetric exchange anisotropy via Dzyaloshinskii-Moriya (DM) interaction that favors canting of in-plane spin order in the ab plane. Specifically, octahedral rotation and tetragonal distortion lead to symmetric and antisymmetric exchange anisotropy terms that are responsible for ϕ-AMR [38, 40]. This is due to the strong SOC that locks the staggered AFM pseudospins moments to the antiferrodistortive octahedral, and as a result, canted pseudospin as well as the net in-plane moment is obtained. To prevent the cancelation of canted moments, they must be aligned parallel to each other across the interlayer. The in-plane canted moments and their stability depend on the strength of interlayer AFM coupling as well. The Hamiltonian is originally proposed for layered Sr_2IrO_4 for explaining the strength of interlayer coupling between IrO_2 layers separated by nonmagnetic SrO layer. In a different study, the theory of interlayer coupling is used to describe the magnetic properties of artificially synthesized SLs, i.e., $SrIrO_3/SrTiO_3$ [41]. Increasing the insulating $SrTiO_3$ layer helps to reduce interlayer coupling and decrease magnetic transition temperature. The magnetic exchange interactions can be governed directly from one layer to another layer of the IrO_2 plane separated by the SrO layer in Sr_2IrO_4 or can be governed by electron hopping through the $SrTiO_3$ layer in $SrIrO_3/SrTiO_3$ SL. As the periodicity increases from $(SrIrO_3)1/(SrTiO_3)1$ (1/1-SL) to $(SrIrO_3)1/(SrTiO_3)2$ (1/2-SL), the transition temperature is significantly reduced due to the decrease in interlayer exchange coupling. The insertion of additional insulating $SrTiO_3$-blocking layers hinders the electronic hopping and exchange coupling along the c-axis, causing a decrease in interlayer coupling. Both Sr_2IrO_4 and 1/1-SL exhibit net magnetizations originating from the canting of the AFM moments within the IrO_2 plane [22]. The coupling of $LaNiO_3$ to the insulating FM LaMnO3 at the interface can lead to the stabilization of an induced AFM order in the [111] direction, which can generate interlayer AFM coupling between two $LaMnO_3$ layers separated by 7 u.c of $LaNiO_3$, as suggested by the authors in a separate study [21].

3.2 $CaMnO_3$ and $CaIrO_3$ periodicity dependent AMR study in $CaMnO_3/CaIrO_3$ SLs

In this section, we discuss a study conducted to understand further the role of individual layers in AMR. The study aims to determine the specific role of individual layers in achieving a large AMR in $(CaIrO_3)x/(CaMnO_3)y$ SLs. The AMR of different sets of heterostructures is analyzed, and the results are presented in **Figure 7 (a-d)**.

i. $(MIx2)_5$ for x = 4, 6, and 8 as $(MI42)_5$, $(MI62)_5$, and $(MI82)_5$, with the $CaIrO_3$ period fixed to 2 u.c.: The ϕ-AMR values of this series of SLs are plotted in **Figure 7(a)**. It is clear that, except for x = 2, the ϕ-AMR of all other SLs is in the range of 0.4–0.8%.

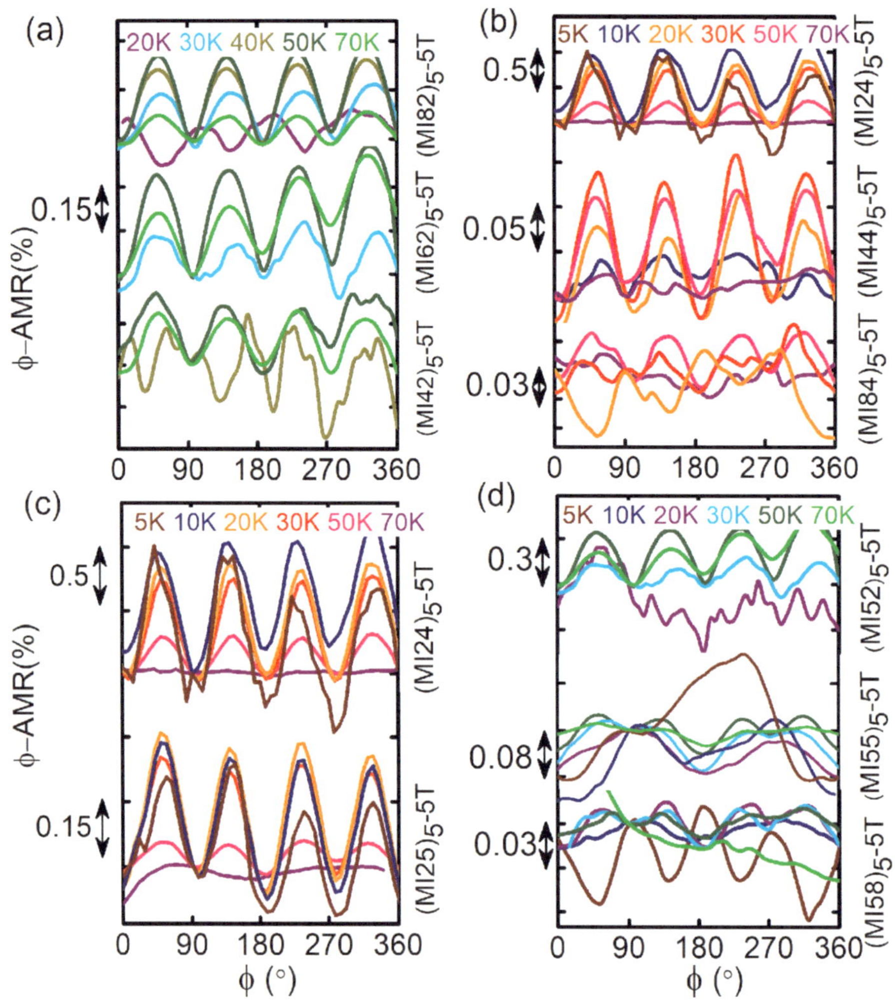

Figure 7.
Exhibits the ϕ-AMR for (a) $(MIx2)_5$ with x = 4, 6, and 8, and (b) $(MIx4)_5$ with x = 2, 4, and 8, recorded at various temperatures with a 5 T magnetic field. Figures (c) and (d) display the ϕ-AMR for (MI2y)z with y = 4 and 5 and (MI5y)z with y = 2, 5, and 8, respectively, measured at different temperatures with a magnetic field of 5 T. (Reprinted from [14] © 2023 American Physical Society).

ii. $(MIx4)_5$ for x = 2, 4, and 8 as $(MI24)_5$, $(MI44)_5$, and $(MI84)_5$, with the $CaIrO_3$ period fixed to 4 u.c.: In this case, the variation of the $CaMnO_3$ period for a fixed larger $CaIrO_3$ period (4 u.c.) shows a ϕ-AMR in the range of ~0.02–1.5%; however, except for $(MI24)_5$, the AMR of the SLs is less than 0.2%. These data are plotted in **Figure 7 (b)**.

iii. The SLs (MI2y)z for y = 4 and 5 are considered, with a fixed $CaMnO_3$ period of 2 u.c. The data shows that as the $CaIrO_3$ period increases from 2 to 4 and 5, the AMR decreases significantly. The plot of the AMR data is given in **Figure 7 (c)**.

iv. The SLs (MI5y)z for y = 2, 5, and 8 are studied, with a fixed $CaMnO_3$ period of 5 u.c. It is observed that the decrease in AMR is pronounced as the $CaIrO_3$

period increases. For $(MI58)_5$, the AMR follows the behavior of a $CaIrO_3$ film of the same thickness. The plot of the AMR data is given in **Figure 7(d)**.

v. It is found that the AMR of $(MI58)_5$, having a large $CaIrO_3$ period of 8 u.c., matches with that of a 10-nm-thick $CaIrO_3$ film in terms of phase and amplitude.

The first three points suggest that a short $CaMnO_3$ period contributes to a high AMR value when the $CaIrO_3$ period is also small. However, when the maximum period of $CaMnO_3$ and $CaIrO_3$ in $CaMnO_3/CaIrO_3$ SLs is reached, the interlayer coupling necessary for biaxial anisotropy is lost, and the SL behaves like a single $CaIrO_3$ film, as stated in the fifth point. SLs with a higher periodic thickness of $CaMnO_3$ and $CaIrO_3$, such as the (MI33)5 and (MI24)5 SLs, exhibit reduced octahedral distortion. As a result, these SLs demonstrate a decrease in AMR response compared to the (MI22)10 SL. This reduction results in a lower possibility of obtaining a net canted moment from each IrO_2 plane, thereby reducing the AMR response of $(MI33)_5$ and $(MI24)_5$ SLs in comparison to the $(MI22)_{10}$ SL. The AMR value of $\sim$1% for $(MI24)_5$ and $(MI52)_5$ SL confirms that the $CaMnO_3$ and $CaIrO_3$ periods have an equal impact on interlayer coupling in $CaMnO_3/CaIrO_3$ SLs. Therefore, in $CaIrO_3$/ $CaMnO_3$ SLs, the similarity of magnetic phase and structural distortion of $(a^-a^-c^+)$ type in low dimensions is a unique attribute and can be argued as a decisive factor for strong interlayer coupling. The structural distortion of both constituent layers in $CaIrO_3/CaMnO_3$ heterostructures contributes to a parallel interlayer alignment of moments, as shown in previous studies [19, 42]. This results in biaxial anisotropy and a large fourfold sinusoidal AMR in the heterostructures.

3.3 Dynamics of ϕ-AMR in (MI22)z: Biaxial anisotropy and spin-flop transition

The trough and crest of ϕ-AMR are assigned by the difference in scattering by soft (100) and hard (110) axes in $(MI22)_{10}$ SL, a biaxial anisotropic system, as shown in **Figure 8 (a, b)**. There appears a transition from uneven scattering from (110) family of axes at 25 K to (100) axes at 14 K, as emphasized in **Figure 8 (c)**.

At 25 K, the stronger crest peaks at (110) and (−1–10) axes suggest larger scattering compared to that at (−110) and (1–10) peaks, despite the uniformity in scattering observed by the (100) family of axes, as shown in **Figure 8 (b, c)**. Similarly, at 14 K, crest peaks are uniform in magnitude whereas trough peaks are not uniform in magnitude. Hence, the ϕ-AMR in $(MI22)_{10}$ SL deviates from regular four-fold sinusoidal symmetry. As scrutinized in very close temperature intervals in the range of 10–25 K, as shown in **Figure 8(c)**, as one moves from the (100) to the (110) crystal axes, a new superimposed feature in the form of a four-fold pattern of AMR kinks emerges alongside the underlying sinusoidal pattern. At 25 K, there is a clear and well-defined four-fold pattern of ϕ dependent AMR, but as the temperature decreases to 22 and 18 K, the pattern becomes more complex, with multiple kinks. Finally, at 15 K, the pattern reverts to a symmetric and sharp four-fold single kink configuration. The smooth pattern appears at 14 K which further transforms to a sharp step-like humungous amplitude of 70% at 10 K. The polarity of AMR peak amplitudes transforms for different fields, i.e., 9 and 5 T, as shown in **Figure 9 (a)**. This unprecedented ϕ-AMR is complex both in its pattern and amplitude. At 5 T, the smooth sinusoidal modulation of the ϕ-AMR displays a modest 10% increase, as shown in **Figure 9 (a)**. However, when the magnetic field is increased to 9 T, a remarkable metamagnetic

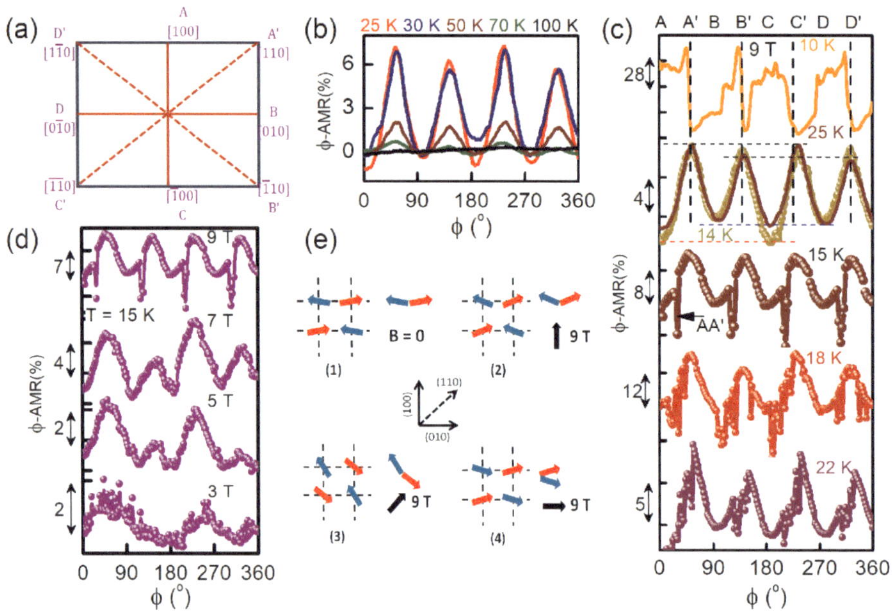

Figure 8.
(a) The crystallographic in-plane directions for the (100) and (110) families are labeled as a, B, C, D, and A′, B′, C′, D′ respectively. (b) The ϕ-AMR for $(MI22)_{10}$ is shown at different temperatures between 25 K and 100 K with a magnetic field of 9 T. (c) The ϕ-AMR for $(MI22)_{10}$ is displayed for the temperature range between 10 K and 22 K, showing a fourfold symmetry, and indicating the onset of the spin-flop transition at 22 K. (d) The field dependence of ϕ-AMR for $(MI22)_{10}$ is presented at 15 K. (e) The spin arrangement in relation to the AMR is shown. The spins depicted in red and blue correspond to the two sublattices of the antiferromagnetic order. A subtle canting in AFM order at B = 0 is observed, which increases at a field of 9 T along the easy (100) axis (1 and 2), while the effect is less pronounced when B is applied along the (110) hard axis (panel 3). For B = 9 T along (010), the spin-flop arrangement at 10 K is depicted in the last panel (panel 4). (Reprinted from [14] © 2023 American Physical Society).

transition occurs, which is responsible for generating a substantial 70% step-like AMR. This transition is identified as a spin-flop transition [43] since the trough of the ϕ-AMR at 5 T coincides with the crest at 9 T. By comparing the ϕ-AMR at these two fields, it is clear that there is an abrupt drop in resistance at 9 T when the sinusoidal ϕ-AMR peak starts appearing at 5 T, as shown in **Figure 9 (a)**. These observations establish that the spin-flop metamagnetic transition is responsible for the enormous ϕ-AMR at 9 T and 10 K. Magnetic-field dependence of ϕ-AMR was conducted at 15 and 25 K to investigate spin-flop-induced transition. **Figure 9 (b)** and **(c)** shows a kink-like transition indicative of the spin-flop transition only at 9 T and 15 K, absent at 25 K. Two primary phenomena are responsible for the diverse characteristics of the ϕ-AMR. Firstly, a robust biaxial magneto-crystalline anisotropy contributes to the sinusoidal ϕ-AMR, which can reach up to 20%. Secondly, the spin-flop transition induces kink- and step-like metamagnetic AMR of up to 70% at a maximum field of 9 T. To explain the anomalous AMR of $CaIrO_3/CaMnO_3$ SLs, a competition between pseudospin–lattice (S-L) coupling and field-pseudospin coupling is proposed. The S-L coupling in iridates is represented by Eq. (3) with the Hamiltonian Hs-ι [19, 38].

$$H_{s-\iota} = \Gamma_{(x^2-y^2)}\ \cos(2\alpha)(S_i^x S_j^x - S_i^y S_j^y) + \Gamma_{(xy)}\ \sin(2\alpha)(S_i^x S_j^y + S_i^y S_j^x) \tag{3}$$

The energy scales of S-L coupling to the distortions along (100) and (110) are denoted by $\Gamma_{(x^2-y^2)}$ and $\Gamma_{(xy)}$, respectively, while α represents the angle between the staggered moments and the (100) axis. As it is mentioned previously that AFM

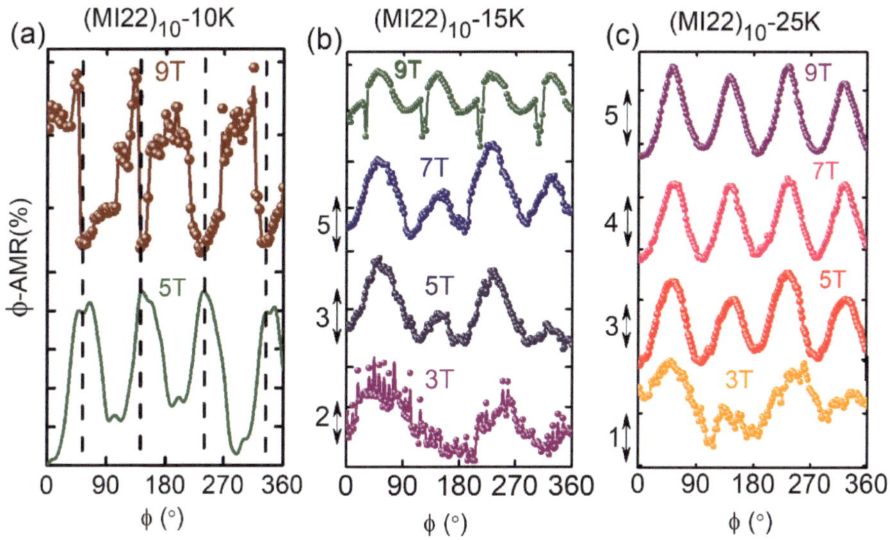

Figure 9.
Comparison of ϕ-AMR at T = 10 K for $(MI22)_{10}$ reveals differences between measurements at 9 and 5 T. field dependence of ϕ-AMR for $(MI22)_{10}$ is measured at 15 and 25 K. (Reprinted from [14] © 2023 American Physical Society).

staggered pseudospins in ab plane are entangled to the antiferrodistortive octahedral via SOC, resulting canted pseudospin, hence, pseudospins have some special symmetric directions, i.e., xy and $x^2 - y^2$ quadruple symmetries. The competition between xy and $x^2 - y^2$ quadruple symmetries of pseudospins have two solutions: α=0 for $\Gamma_{(x^2-y^2)} < \Gamma_{(xy)}$ and α=45° for $\Gamma_{(x^2-y^2)} > \Gamma_{(xy)}$.$Sr_2$ IrO_4 and an engineered SL, $SrIrO_3$/$SrTiO_3$, both possess a similar magnetic structure and exhibit the former phenomenon, whereas the latter is observed in $CaIrO_3$/$SrTiO_3$ heterostructures. In both cases of $SrIrO_3$ and Sr_2IrO_4, the ϕ-AMR phase lags by 45° compared with $CaIrO_3$/$CaMnO_3$ SL [19, 44]. In the case of $CaIrO_3$/$CaMnO_3$ SLs, the optimal solution is achieved at α=0 since the minimum of AMR oscillation aligns with the (100) axes, which is analogous to that of CaIrO3/SrTiO3 SLs. The difference in the phase lag between $SrIrO_3$- and $CaIrO_3$-based 3d-5d SLs is due to the different sense of octahedral rotations in their low-dimensiona limits (**Figure 10**).

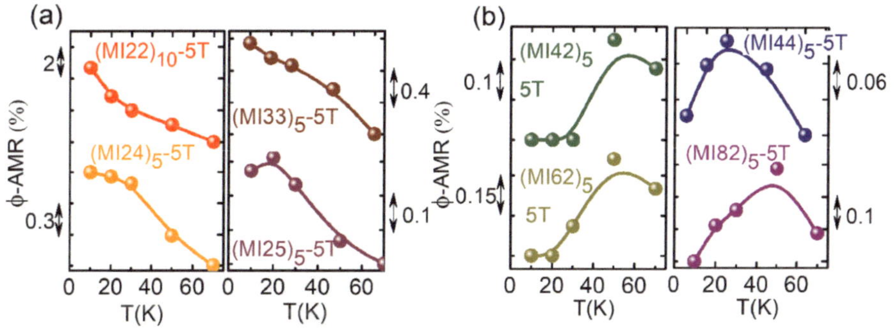

Figure 10.
Temperature dependence of the ϕ-AMR amplitude for (MIxy)z SLs at H = 5 T. (Reprinted from [14] © 2023 American Physical Society).

3.4 Temperature dependence of ϕ-AMR in the context of magneto-elastic coupling

The behavior of solid-state materials in the vicinity of a magnetic transition is a complex and fascinating area of research. At such points, the coupling between the spin and lattice degrees of freedom, known as S-L coupling, plays a critical role, and its strength changes as the temperature decreases [38, 45]. One of the most striking manifestations of S-L coupling is the appearance of a sinusoidal ϕ-AMR due to the continuous lattice deformation under the influence of an external magnetic field. There are two ways in which S-L coupling can occur in $CaMnO_3/CaIrO_3$ SLs, namely, the coupling of pseudospins with octahedral distortion and the response of pseudospins to lattice vibrations or phonons that vary with temperature. It is noteworthy that the structural transition responsible for octahedral deformations has not been reported yet in layered systems, i.e., $SrIrO_3/SrTiO_3$ or $CaIrO_3/SrTiO_3$ SL.

As the temperature changes, the strength of S-L coupling can be described in two regimes based on the magnetic moment of the layered system. (i) In lower magnetic moment-based SLs such as $(MIx2)_5$ (x = 4, 5, 8), the anisotropic scattering of electrons due to the orbital degree of freedom via lattice vibration is more prominent near the transition temperature, leading to strong pseudospin-lattice coupling. At lower temperatures, however, due to the stiffness of the lattice, the response of pseudospins with the lattice is low, and AMR is expected to be governed by field-pseudospin coupling only. (ii) In higher magnetic moment-based SL such as $(MI22)_{10}$, at a temperature of 10 K, the $(MI22)_{10}$ SL exhibits a sharp four-fold single kink separated by 90° with no sinusoidal variation observed in its ϕ-AMR. This suggests that the orbital degree of freedom of electrons contributed by lattice vibration is suppressed at this temperature, and the step-like AMR pattern in $(MI22)_{10}$ SL is a result of field-pseudospin coupling, facilitated by higher magnetic moment. In contrast, the absence of the kink pattern in $(MIx2)_5$ (x = 4, 5, 8) represents the absence of field-pseudospin coupling. Thus, the AMR mechanism in $(MIx2)_5$ (x = 4, 5, 8) SLs is mainly governed by orthorhombic distortion through S-L coupling restoring the sinusoidal resistance [22].

Furthermore, it is noteworthy that the sinusoidal ϕ-AMR pattern with a kink observed in the temperature range of 15 to 22 K in the $(MI22)_{10}$ SL indicates the coexistence of both S-L coupling and field pseudospin coupling. Additionally, in the cases of both $(MI22)_{10}$ and $(MI33)_5$ SLs, the amplitude of ϕ-AMR steadily increases with decreasing temperature. This suggests that the competition between these two couplings, which varies with temperature, ultimately determines the ϕ-AMR behavior in the $(MI22)_{10}$ SL. Specifically, at high temperatures, the S-L coupling is responsible for the reorientation of moments, whereas at lower temperatures when the lattice is rigid but the moments are larger, the direct coupling of field pseudospin dominates.

3.5 Metamagnetic transitions in perovskite manganites and iridates

Metamagnetic transitions are a fascinating phenomenon that has been observed in a variety of materials including manganites and iridates. Over the past few decades, much attention has been focused on the ABO_3 perovskite-type mixed valent manganites which have the general formula, i.e., $R_{1-x}A_xMnO_3$, where 'R' represents a trivalent rare-earth cation and 'A' represents a divalent cation [46]. These materials have been extensively studied due to their intriguing properties arising from the interplay between the charge-orbital coupling. One of the most interesting features of these compounds is the colossal magnetoresistance phase that emerges due to the interplay between charge and orbital ordering. The evolution of different magnetic ground states is also observed in these compounds as the value of 'x' varies from 0 to 1. For instance, in $Nd_{1\text{-}x}Sr_xMnO_3$, the magnetic phase transitions from FM metallic state to a C-type insulating phase as 'x' increases from 0.3 to 0.8, via an A-type AFM metallic phase with a ferromagnetic plane featuring uniform $d_{(x^2-y^2)}$-type orbital order. In another material, $Pr_{1\text{-}x}Ca_xMnO_3$, a transition from a charge-orbital ordered AFM-CE type insulating state to FM metallic state is observed as the temperature decreases in the presence of an external magnetic field. Moreover, a larger field of 27 T is required to destabilize the charge-ordered insulating state via a metamagnetic transition at low temperatures in the half-doped $Pr_{1\text{-}x}Ca_xMnO_3$ ($x = 0.5$) material [46, 47]. These results indicate that mixed valent manganites exhibit strong spin-charge-orbital-lattice coupling, resulting in the field-induced metamagnetic transition [48]. More recently, materials based on late transition metal ions with strong SOC have come into focus. These compounds no longer treat spin-orbital separation as a separate entity that requires magnetism to be reformulated using pseudospin. Spin-flop-based metamagnetic transitions, which are the result of spin-lattice coupling, have been observed in these materials [45, 49]. As an example, the theoretical basis for the spin-flop transition lies in the stronger interlayer coupling that varies from Sr_2IrO_4 to $Sr_3Ir_2O_7$. Overall, the occurrence of metamagnetic transitions in manganites and iridates, as well as the diverse phenomena associated with these transitions, highlight the complex interplay between charge, orbital, and spin degrees of freedom in these materials.

3.5.1 Metamagnetic transition in thin-film

Obtaining a spin-flop transition in thin films can pose a significant challenge due to the reduced thickness of the film. However, in a thin film, the magnetic moments are often constrained by the surface and interface effects as well as defects [48, 50–53]. These effects can create a preferential orientation for the magnetic moments, hindering their ability to rotate easily in response to an external magnetic field. It is crucial to have a strong SOC effect to effectively leverage large magnetoresistance anisotropy. However, if the magnetic moments are aligned along an easy axis that is perpendicular to the film plane, it becomes quite challenging to rotate them away from this easy axis and towards the basal plane. As a result, utilizing TAMR effects may be impossible. On the other hand, in Mn_2Au, the magnetic moments are oriented in the basal plane of the structure. This makes it relatively easy to rotate them from one easy in-plane direction to another with the aid of a relatively small external excitation [54]. Defects and imperfections in thin films can also have a considerable impact on the spin-flop transition based on the rotation of spin [55]. These defects can act as pinning sites for the magnetic moments, impeding their motion. Thus, a sharp interface is essential for achieving a spin-flop transition. Careful consideration of the film thickness, interface effects, shape anisotropy, and defects, among other factors, is necessary to optimize the magnetic properties and achieve the desired behaviors. The resistance oscillations in the current $CaIrO_3/CaMnO_3$ SLs stem from a magnetic moment that oscillates with respect to the crystallographic axis embedded in a system with in-plane biaxial magnetic anisotropy. Along with these oscillations, there are sharp kink- and steplike transitions that induce an additional large component ϕ-AMR and originate from spin-flop metamagnetic transition. In **Figure 8 (e)**, there is a depiction of a pseudospin arrangement that corresponds to the crest and trough of the spin-flop-based AMR. When a field of 9 T is applied along the (100) easy axis, the canting angle and magnetic moment increase. The canting effect is less pronounced when the field is applied along the (110) hard axes, where the AFM spin arrangement is rotated along the direction of the field. When the magnetic moments are tilted at larger angles along the (100) and (010) directions, the material shows less resistance to the flow of electric current compared to when the magnetic moments are tilted along the (110) direction. Additionally, **Figure 8 (e)** illustrates a spin-flop transition that occurs at 10 K in a field of 9 T.

4. Conclusion

Ultimately, it should be highlighted that the selection of a suitable 3d compound is crucial for achieving efficient interlayer coupling and distortion, which are necessary to adjust a significant ϕ-AMR. In the case of $CaIrO_3/SrTiO_3$ SLs, the $CaIrO_3$ layers with distorted ($a^-a^-c^+$) octahedral as per Glazer notations convey this distortion between them via distortion of the mediating $SrTiO_3$ layer octahedra, only for one layer thickness of the latter [19]. In present $CaIrO_3/CaMnO_3$ SLs, in contrast, the presence of AMR in (MI82)5 SL, 8 u.c.s thick CaMnO3 mediating SL, suggests that $CaMnO_3$ is the most suitable candidate known so far to promote interlayer exchange coupling. Here, the coupling is such long-range that the $CaIrO_3$ layers separated by even eight u.c.s of $CaMnO_3$ in $(MI82)_5$ SL are capable of inducing a ϕ-AMR of 0.25% at 5 T and 50 K. The reason for this is that when few unit cells are considered, both $CaMnO_3$ and $CaIrO_3$ films exhibit orthorhombic distortion with the same octahedral rotation pattern ($a^-a^-c^+$) according to Glazer's notations. Moreover, they also share a similar in-plane DM-type canted AFM phase. This similarity of distortion of both the constituents is key for phenomenal $CaIrO_3$ interlayer coupling, and, hence, a large biaxial anisotropy is obtained. The presence of a stronger H_{EB}, as observed in (MI44)5 and (MI84)5, leads to the formation of a graded AFM/FM phase within the $CaMnO_3$ layer, which acts as a hindrance to a uniform interlayer coupling of $CaIrO_3$. The impact is evident in the comparison between (MI22)10, which has an AMR of 23% in the absence of H_{EB}, and (MI33)5, which has a moderate H_{EB} and a reduced AMR of just 3%.

The $CaIrO_3/CaMnO_3$ SLs prove to be the most potent 3d-5d heterostructures for achieving an unprecedented AMR of about 70%, utilizing two key factors of a strong biaxial anisotropy and a spin-flop metamagnetic transition. On the fundamental side, employing the tolerance factor to appropriately manage the structural and surface layer construction of a large bi-axial magnetic anisotropy, fine-tuning the interlayer coupling facilitated by an exceptionally thick layer, and showcasing the spin-flop transition to enhance the degree of anisotropic AMR are all significant advancements in the realm of 3d-5d SLs. These developments hold immense promise in the field of contemporary quantum materials and their application in technological advancements. This proof-of-concept study opens new avenues for designing highly sensitive AMR readout devices for emerging AFM spintronics.

Acknowledgements

The authors extend their gratitude to several organizations for their financial support and provision of research facilities. Specifically, D.S.R. expresses thanks to the Department of Science and Technology (DST) Nanomission and the Science and

Engineering Research Board Technology, New Delhi for their financial support through Research Project No. SM/NM/NS-84/2016 and Project No. CRG/2020/002338, respectively. XAS measurements were performed at UC San Diego as part of a search for materials for spin-torque oscillators, which was supported by Quantum Materials for Energy Efficient Neuromorphic Computing, an Energy Frontier Research Center funded by the U.S. Department of Energy (DOE), Office of Science, Basic Energy Sciences under Award No. DE-SC0019273. Also, Resources for this research were made available through the Advanced Photon Source, which is operated by Argonne National Laboratory for the DOE Office of Science under Contract No. DE-AC02-06CH11357, with supplementary support from the National Science Foundation under Grant No. DMR-0703406. The authors also acknowledge the DOE Office of Science for supporting extraordinary facility operations through the National Virtual Biotechnology Laboratory. The authors express their appreciation to various individuals for their assistance in performing magnetization, transport, and x-ray diffraction characterizations, as well as in the preparation of transmission electron microscopy specimens. Finally, the authors express their gratitude for the utilization of the HZDR Ion Beam Center TEM facilities and the financial support provided by the German Federal Ministry of Education and Research (Grant No. 03SF0451) through the Helmholtz Energy Materials Characterization Platform for the funding of TEM Talos.

Author details

Suman Sardar
Department of Physics, Garhbeta College, India

*Address all correspondence to: sumansardar@garhbetacollege.ac.in

References

[1] Němec P, Fiebig M, Kampfrath T, Kimel AV. Antiferromagnetic opto-spintronics. Nature Physics. 2018;**14**: 229-241

[2] Šmejkal L, Mokrousov Y, Yan B, MacDonald AH. Topological antiferromagnetic spintronics. Nature Physics. 2018;**14**:242-251

[3] Jungwirth T et al. The multiple directions of antiferromagnetic spintronics. Nature Physics. 2018;**14**: 200-203

[4] Jungfleisch MB, Zhang W, Hoffmann A. Perspectives of antiferromagnetic spintronics. Physics Letters A. 2018;**382**:865-871

[5] Baltz V et al. Antiferromagnetic spintronics. Reviews of Modern Physics. 2018;**90**:15005

[6] Jungwirth T, Marti X, Wadley P, Wunderlich J. Antiferromagnetic spintronics. Nature Nanotechnology. 2016;**11**:231-241

[7] Stamps RL. Mechanisms for exchange bias in the UK mechanisms for exchange bias. Journal of Physics D: Applied Physics. 2000;**33**:247-268

[8] Nogués J, Schuller IK. Exchange bias. Journal of Magnetism and Magnetic Materials. 1999;**192**:203-232

[9] Rana R, Pandey P, Singh RP, Rana DS. Positive exchange-bias and giant vertical hysteretic shift in La 0.3 Sr0.7 FeO3/SrRuO3 bilayers. Scientific Reports. 2014;**4**:1-8

[10] Železný J et al. Relativistic néel-order fields induced by electrical current in antiferromagnets. Physical Review Letters. 2014;**113**:1-5

[11] Park BG et al. A spin-valve-like magnetoresistance of an antiferromagnet-based tunnel junction. Nature Materials. 2011;**10**: 347-351

[12] Avci CO et al. Current-induced switching in a magnetic insulator. Nature Materials. 2017;**16**:309-314

[13] Huang K et al. Tailoring magnetic order via atomically stacking 3 d/5 d electrons to achieve high-performance spintronic devices. Applied Physics Reviews. 2020;**7**:011401-1-011401-7

[14] Vagadia M et al. Extraordinary anisotropic magnetoresistance in $CaMnO_3/CaIrO_3$ heterostructures. Physical Review B. 2022;**105**:1-7

[15] Yan H et al. Electric-field-controlled antiferromagnetic spintronic devices. Advanced Materials. 2020;**32**:1-23

[16] Egilmez M, Chow KH, Jung JA. Anisotropic magnetoresistance in perovskite manganites. Modern Physics Letters B. 2011;**25**:697-722

[17] Nam Hai P, Sasaki D, Anh LD, Tanaka M. Crystalline anisotropic magnetoresistance with two-fold and eight-fold symmetry in (In,Fe)As ferromagnetic semiconductor. Applied Physics Letters. 2012;**100**:1-6

[18] Bovenzi N, Diez M. Semiclassical theory of anisotropic transport at LaAlO3/SrTiO3 interfaces under an in-plane magnetic field. Physical Review B. 2017;**95**:1-14

[19] Lim ZS et al. Magnetic anisotropy of a quasi two-dimensional canted Antiferromagnet. Nano Letters. 2020;**20**: 1890-1895

[20] Narayanapillai K et al. Current-driven spin orbit field in $LaAlO_3/SrTiO_3$ heterostructures. Applied Physics Letters. 2014;**105**(4):162405-1-162405-4

[21] Gibert M et al. Interlayer coupling through a dimensionality-induced magnetic state. Nature Communications. 2016;7:1-7

[22] Hao L et al. Two-dimensional Jeff = 1/2 antiferromagnetic insulator unraveled from interlayer exchange coupling in artificial perovskite iridate superlattices. Physical Review Letters. 2017;**119**:1-6

[23] Cui Z et al. Correlation-driven eightfold magnetic anisotropy in a two-dimensional oxide monolayer. Science Advances. 2020;**6**:eaay0114 1-8

[24] Biswas A, Jeong YH. Persistent semi-metal-like nature of epitaxial perovskite CaIrO3 thin films. Journal of Applied Physics. 2015;**117**:1-6

[25] Glazer AM. Simple ways of determining perovskite structures. Acta Crystallographica Section A. 1975;**31**: 756-762

[26] Glazer AM. The classification of tilted octahedra in perovskites. Acta Crystallographica Section B Structural Crystallography and Crystal Chemistry. 1972;**28**:3384-3392

[27] Nanda BRK, Satpathy S, Springborg MS. Electron leakage and double-exchange ferromagnetism at the interface between a metal and an antiferromagnetic insulator: CaRuO3/ CaMnO3. Physical Review Letters. 2007; **98**:1-4

[28] Venkatesan T, Wu XD, Muenchausen R, Pique A. Pulsed laser deposition: Future directions. MRS Bulletin. 1992;**17**:54-58

[29] Yi D et al. Atomic-scale control of magnetic anisotropy via novel spin–orbit coupling effect in $La_{2/3}Sr_{1/3}MnO_3/SrIrO_3$ superlattices. Proceedings of the National Academy of Sciences. 2016;**113**: 6397-6402

[30] Connell JG, Nichols J, Gruenewald JH, Kim DW, Seo SSA. Conducting LaAlO3/SrTiO3 heterointerfaces on atomically-flat substrates prepared by deionized-water. Scientific Reports. 2016;**6**:2-7

[31] Lim ZS et al. Emergent topological hall effect at a charge-transfer Interface. Small. 2020;**16**:1-9

[32] Vistoli L et al. Giant topological hall effect in correlated oxide thin films. Nature Physics. 2019;**15**:67-72

[33] He C, Grutter AJ, Gu M, Browning ND, Takamura Y, Kirby BJ, et al. Interfacial ferromagnetism and exchange bias inCaRuO3/ CaMnO3Superlattices. Physical Review Letters. 2012;**109**:197202

[34] Vélez S et al. Hanle magnetoresistance in thin metal films with strong spin-orbit coupling. Physical Review Letters. 2016;**116**:1-6

[35] Peramaiyan G, Sankar R, Muthuselvam IP, Lee WL. Anisotropic magnetotransport and extremely large magnetoresistance in NbAs2 single crystals. Scientific Reports. 2018;**8**:2-9

[36] Suraj TS et al. Tunable and enhanced Rashba spin-orbit coupling in iridate-manganite heterostructures. Physical Review B. 2020;**102**:125145

[37] Nakayama H et al. Spin hall magnetoresistance induced by a nonequilibrium proximity effect. Physical Review Letters. 2013;**110**:1-5

[38] Porras J et al. Pseudospin-lattice coupling in the spin-orbit Mott insulator Sr2IrO4. Physical Review B. 2019;**99**: 1-11

[39] Kim BJ et al. Novel jeff=1/2 Mott state induced by relativistic spin-orbit coupling in Sr2IrO4. Physical Review Letters. 2008;**101**:1-4

[40] Gim Y et al. Isotropic and anisotropic regimes of the field-dependent spin dynamics in Sr2IrO4: Raman scattering studies. Physical Review B. 2016;**93**:1-7

[41] Matsuno J et al. Engineering a Spin-Orbital Magnetic Insulator by Tailoring Superlattices. Physical Review Letters. 2015;**114**:247209-1-247209-5

[42] Rondinelli JM, Spaldin NA. Structure and properties of functional oxide thin films: Insights from electronic-structure calculations. Advanced Materials. 2011; **23**:3363-3381

[43] Machado FLA et al. Spin-flop transition in the easy-plane antiferromagnet nickel oxide. Physical Review B. 2017;**95**:1-10

[44] Fina I et al. Anisotropic magnetoresistance in an antiferromagnetic semiconductor. Nature Communications. 2014;**5**:1-7

[45] Liu H, Khaliullin G. Pseudo-Jahn-teller effect and Magnetoelastic coupling in spin-orbit Mott insulators. Physical Review Letters. 2019;**122**:57203

[46] Dagotto E, Hotta T, Moreo A. Colossal magnetoresistant materials: The key role of phase separation. Physics Reports. 2001;**344**:1-153

[47] Hardy V, Maignan A, Hébert S, Martin C. Calorimetric and magnetic investigations of the metamagnet (formula presented). Physical Review B - Condensed Matter and Materials Physics. 2003;**67**:1-7

[48] Rana DS, Kuberkar DG, Malik SK. Field-induced abrupt change in magnetization of the manganite compounds (LaR)0.45 (CaSr)0.55 Mn O3 (R=Eu and Tb). Physical Review B - Condensed Matter and Materials Physics. 2006;**73**:2-7

[49] Kim JW et al. Dimensionality driven spin-flop transition in layered iridates. Physical Review Letters. 2012;**109**:1-5

[50] Veremchuk I et al. Defect nanostructure and its impact on magnetism of α-Cr_2O_3 thin films. Small. 2022;**18**:1-10

[51] Trallori L. Field and Surface effects on the Ground State of Antiferromagnetic Systems. Physical Review Letters. 1994;**72**:1925-1928

[52] Dantas AL, Carriço AS. Surface-induced low-field instability of antiferromagnetic multilayers. Physical Review B - Condensed Matter and Materials Physics. 1999;**59**: 1223-1231

[53] Rößler UK, Bogdanov AN. Reorientation in antiferromagnetic multilayers: Spin-flop transition and surface effects. Physica Status Solidi C Conferernce. 2004;**1**:3297-3305

[54] Barthem VMTS, Colin CV, Mayaffre H, Julien MH, Givord D. Revealing the properties of Mn 2 Au for antiferromagnetic spintronics. Nature Communications. 2013;**4**(3892):1-7

[55] Qian G et al. Spin-flop transition and Zeeman effect of defect-localized bound states in the antiferromagnetic topological insulator MnBi2Te4. Nano Research. 2023;**16**:1101-1106

Chapter 6

Growth and Interfacial Emergent Properties of Complex Oxide Thin Film Heterostructures

Snehal Mandal

Abstract

Non-trivial/chiral spin textures like skyrmions originate from inversion symmetry breaking. Moreover, inversion symmetry breaking combined with strong spin-orbit coupling (SOC) can lead to a large Dzyaloshinskii-Moriya interaction (DMI). Electrically, these phenomena can be detected through what is called the topological Hall effect (THE). In artificially layered complex oxide thin film heterostructures composed of ferromagnetic or antiferromagnetic layers, this THE appears as an emergent property at the interfaces because it is not intrinsic to the bulk layer of such oxides. Thus these heterostructures provide a playground for the competition among DMI, exchange interaction, and magnetic anisotropy to produce novel non-coplanar spin textures and THE in a designable way due to inversion symmetry breaking at the interfaces. With the advancement in modern fabrication techniques, these properties can be tuned at will by engineering the interfaces of the heterostructures, especially due to crystal structure compatibility of these materials. In this chapter, growth, detection and manipulation of interfacial emergent phenomena in complex oxide heterostructures will be discussed.

Keywords: complex oxide heterostructures, interfaces, magnetism, topological Hall effect, Dzyaloshinskii-Moriya interaction

1. Introduction

Magnetism and magnetic materials lie at the heart of all the electronic storage devices that we all know of (like, magnetic random access memory (MRAM), hard-disk drives (HDD), pen-drives). However, the ever-shrinking dimensions and ever-increasing demands in the storage areal density limit the data transfer rate in the present-day magnetic materials based devices.

Magnetic skyrmions (or simply, skyrmions), which are nanoscale swirling or chiral spin textures arising out of non-trivial real-space topology, are argued to have the potential to be the basis for next-generation magnetic storage devices. Because of the topological protection, these spin textures cannot be unwound without forming a discontinuity, providing them with high stability against external perturbations, even at small sizes. Along with skyrmions, there are other various chiral topological textures that are stabilized due to various magnetic interactions and can be classified by their unique topological properties. The basic requirement for such textures is to have strong spin-orbit coupling (SOC) along with inversion symmetry breaking, which gives rise to Dzyaloshinskii-Moriya interaction (DMI) [1, 2]. Interestingly, this DMI shows up as an anomalous topological contribution to the Hall effect, which thus is called the topological Hall effect (THE), and provides a tell-tale sign of the existence of chiral textures. Although many direct imaging techniques have been developed over the last few years, electrical detection by means of THE has also become a promising technique.

To harness topological textures practically, it is important to develop materials that can stabilize these textures across a wide range of temperatures, and develop all-electrical pathways to manipulate them. A large family of topological textures and control mechanisms have been discovered in magnetic metal-based heterostructures, or chiral magnets such as Heusler compounds [3] and B_{20} systems [4, 5], making them favorites for developing skyrmionics. However, to no wonder, in nature only a few bulk materials have crystal structures that intrinsically aids to the formation of skyrmions. One way out is to design systems in such a manner that these phenomena can "emerge" due to some new interactions, which otherwise are absent in the bulk. Thin films offer this extra space where one can utilize the interfaces and surfaces to modify/tune the interactions separately from their bulk counterparts. Most often, these phenomena emerge at or near the surfaces/interfaces of thin films/heterostructures that are designed at will; due to which they are called "interfacial emergent phenomena".

Correlated oxide magnets have sparked great attention because they provide the following specific advantages: **(i)** They host myriad of phases (like, magnetic, multiferroic, *etc.*) due to strong correlations among various degrees-of-freedom (*i.e.*, charge, spin, orbital and lattice), which are quite susceptible to external perturbations. These couplings provide innovative practical avenues for controlling both intrinsic and emergent magnetic characteristics, such as anisotropy, symmetric exchange interactions, and DMI [6]. **(ii)** Because of their lower and configurable charge carrier densities, oxides are very responsive to electric fields, making them excellent candidates for non-volatile control. **(iii)** Additionally, using standard growth techniques, high-quality crystalline oxide thin films with carefully regulated interfaces and/or surfaces and heterostructures thereof, permitting spatial inversion symmetry breaking (with extremely low defects), can be fabricated at will. This chapter focuses on the two key proponents in the field of oxide skyrmionics: the creation of oxide heterostructures and detection and tuning mechanisms, which might set the course for practical oxide-based devices.

We start with a brief discussion on the basic physical interactions that can generate non-collinear or chiral spin textures, followed by their notions in electrical/magnetotransport properties. As a prelude to the electrical detection of such interactions/spin textures, we then discuss the various types of Hall effects and their phenomenological origin that show up in the experiments and cover a large part of the chapter on how to extract (from raw data) the topological Hall effect (THE), *i.e.* associated with such chiral spin structures. Finally, we discuss the ways one can harness these properties in complex oxide heterostructures. We go straight into the growth/development part of complex oxide heterostructures, followed by the detection and manipulation of the interfacial emergent properties in them. The recent progress in this field of research along with some future outlook is also concluded.

2. Brief basics of Skyrmions

2.1 Skyrmions

A magnetic skyrmion (or simply skyrmion) is a collection of magnetic moments forming non-coplanar/chiral texture. Mathematically it can be described by the topological skyrmion number (N_{sk}), which counts the times magnetic moments wraps a unit sphere, and is given by the following expression:

$$N_{sk} = \frac{1}{4\pi} \iint \vec{m}\left(\vec{r}\right) \cdot \left(\frac{\partial \vec{m}\left(\vec{r}\right)}{\partial x} \times \frac{\partial \vec{m}\left(\vec{r}\right)}{\partial y} \right) d^2\vec{r} \tag{1}$$

where $\vec{m}(\vec{r})$ is the magnetic moment. For trivial spin arrangements/textures, like ferromagnet or antiferromagnet, $N_{sk} = 0$. For magnetic vortices (as in magnetic nanostructures), $N_{sk} = \pm\frac{1}{2}$. For non-trivial spin textures like skyrmions, $N_{sk} = 1$. There are two basic variations of skyrmions with different spin arrangements along the radial direction, *viz.*, Bloch-type skyrmions and Néel-type skyrmions. In Bloch-type skyrmions, the spins rotate in the tangential planes (that is perpendicular to the radial directions), when moving from the Center to the circumference (periphery). Whereas, in a Néel-type skyrmions, the spins rotate in the radial planes when moving from the Center to the circumference. These are schematically shown in **Figure 1**. It is worth mentioning that there are other types of chiral spin textures apart from skyrmions and the skyrmion number topological charge can be used to identify the topological distinction of these different types of spin textures.

The type of chiral spin texture formation depends on the interplay of different magnetic interactions in a material.

2.2 Interactions involved in stabilizing chiral spin textures

There are various magnetic interactions that can generate skyrmions in magnetic systems, and at times, multiple such mechanisms may contribute simultaneously.

2.2.1 Dzyaloshinskii-Moriya interaction (DMI)

The Dzyaloshinskii-Moriya interaction (DMI) is an asymmetric exchange interaction that favors canting of spins in materials that would, in contrary, be ferromagnetic (FM) or antiferromagnetic (AF) with collinearly aligned spins (either parallel or anti-

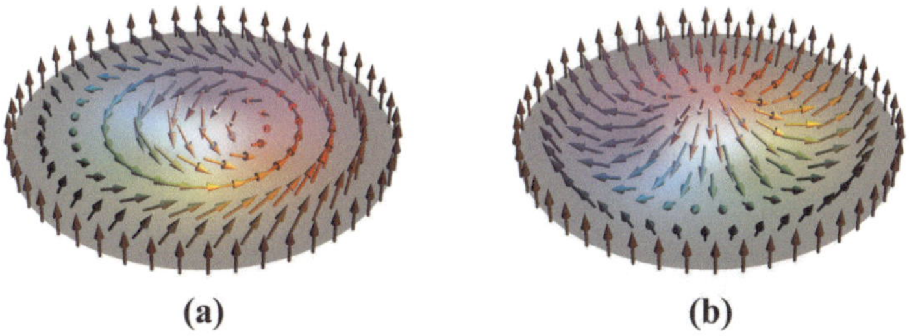

Figure 1.
Schematic of spin textures in two types of skyrmions: (a) Bloch-type skyrmion, (b) Néel-type skyrmion.

parallel). The DMI originates from spin-orbit coupling (SOC) and inversion symmetry breaking (typically through bulk crystal structure or multilayer heterostructure). Its Hamiltonian is written as:

$$H_{DMI} = \sum_{i,j} \vec{D}_{ij} \cdot \left(\vec{S}_i \times \vec{S}_j \right) \tag{2}$$

where $\vec{D}_{ij}$ is the DMI vector. The direction of the DMI vector depends on the symmetry of the system. For bulk material with no inversion symmetry such as B_{20} compounds, the $\vec{D}_{ij}$ is parallel to the vector that connects $\vec{S}_i$ and $\vec{S}_j$ [4, 7]. On the other hand, in multilayer heterostructure composed of a FM layer and a layer with strong SOC, the $\vec{D}_{ij}$ is often parallel to the interface [8], as shown in **Figure 2**.

2.2.2 Magnetic dipolar interaction

In magnetic thin films with perpendicular easy-axis anisotropy, the dipolar interaction favors an in-plane magnetization, whereas the anisotropy prefers an out-of-plane magnetization. The competition between these two interactions results in periodic stripes in which the magnetization rotates in the plane perpendicular to the thin film. An applied magnetic field perpendicular to the film turns the stripe state into a periodic array of magnetic bubbles or skyrmions.

Apart from these, **frustrated exchange interactions** and **four-spin exchange interactions** can also aid to the formation of chiral spin structures. However, in these two cases, the skyrmions are of atomic size length scales (often of the order of the lattice constant ~1 nm).

In case of dipolar interactions, skyrmions are typically of the order of 100 nm to 1 μm, which is comparable to the period of the spiral determined by the ratio of the dipolar and exchange interactions. In case of DMI, the size is determined by the strength of DMI and is typically 5 nm – 100 nm. As a result, skyrmions in skyrmion crystals in cases (1) and (2) are larger than the lattice constant, and therefore the continuum approximation is justified. In these two cases, the energy density of the skyrmions is much smaller than the atomic exchange energy J, and the topological protectorate holds. In other words, discontinuous spin configurations — called monopoles — with energy of the order of J can create or annihilate the skyrmions.

In complex oxide based thin film heterostructures, one can easily comply with the dipole exchange interaction as well as the DMI interaction by harnessing the inversion

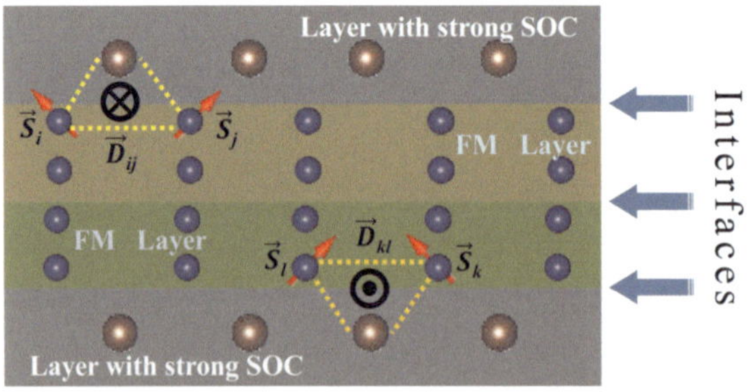

Figure 2.
Schematic of the interfacial DMI mechanism that gives rise to emergent phenomena in thin film heterostructures.

symmetry breaking at the interfaces and/or surfaces of the heterostructures, using heavy Z-based elements (Z = atomic number) for strong SOC and of course the thickness to generate anisotropy along various directions with respect to the film plane.

2.3 Detection techniques

There are many various techniques for detection of skyrmions. These include advanced imaging techniques like magnetic force microscopy (MFM), Lorentz transmission electron microscopy (LTEM), spin polarized scanning tunneling microscopy (SP-STM), photoemission electron microscopy (PEEM), *etc.*; synchrotron radiation based techniques like x-ray magnetic circular dichroism (XMCD). But very intriguing and easy technique is the electrical one, the topological Hall effect (THE). Electrical detection can be easily performed nowadays with the available cryogenic and magnet based systems in any condensed matter physics laboratory.

3. Magnetotransport properties: the family of Hall effect

The Hall effect was discovered by Edwin Hall in 1879 (even before the discovery of the electron) when he observed the evolution of a transverse voltage in conductors for applied electric and magnetic fields which were mutually perpendicular (schematically shown in **Figure 3(a)**) [9]. Couple of years later, he again reported that the effect was ten times larger in ferromagnetic conductors than in non-magnetic conductors [10], which was hence termed as the "Anomalous" Hall effect (AHE). Then, about a century later, other members started to appear in the family, like the spin Hall effect (SHE), quantum Hall effect (QHE), planar Hall effect (PHE), quantum spin Hall effect (QSHE), to name a few. A recent addition to the family is the topological Hall effect (THE) that is related to the chiral spin textures in materials. All of these effects have different origin in themselves, except for the fact that the experimental arrangements remain same (and hence this arrangement is commonly called the "Hall configuration"). For simplicity and distinguishablity, the original Hall effect is often

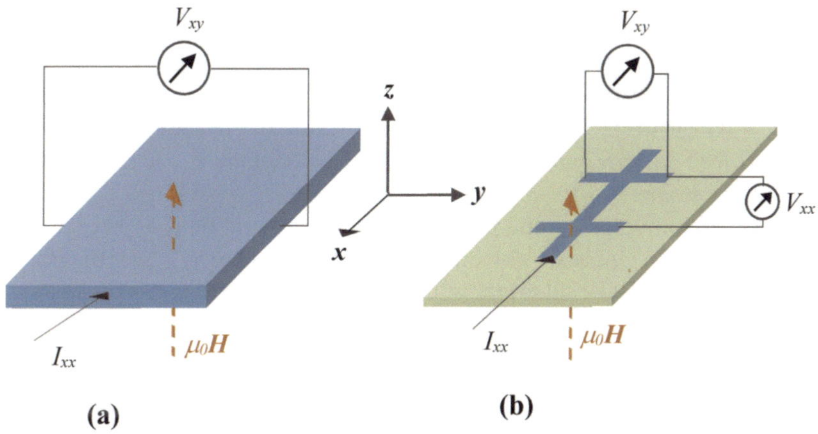

Figure 3.
Basic configuration for Hall effect measurement: (a) for bulk or layered (thin film) samples, (b) for patterned thin film samples in the form of "Hall bar".

called as the ordinary Hall effect (OHE). The schematic of the Hall configuration is shown in **Figure 3**.

Needless to say that this section can form a chapter in itself; so we will try to keep it simple for the readers and discuss phenomenologically some of the important Hall effects, *i.e.*, OHE, AHE, and ofcourse THE, those are of interest for this chapter.

3.1 Ordinary Hall effect (OHE)

When electric current (I) flows (say along x-direction) through a metal or a semiconductor placed in a perpendicular magnetic field ($\mu_0 H$, say along z-direction), the charge carriers inside the material are deflected from the straight path by the magnetic field as per the Lorentz force (given by Eq. (3)).

$$\vec{F} = q\left[\vec{E} + \left(\vec{v} \times \vec{B}\right)\right] \tag{3}$$

This results in charge accumulation on the sides (in y-direction) preferably depending on the type of charge carrier, giving rise to a transverse voltage along the y-direction (*i.e.*, mutually perpendicular to that of both I and $\mu_0 H$). This voltage is called the Hall voltage (V_{xy}); and in a non-magnetic material it is proportional to the applied magnetic field. The Hall resistance ($R_x y = V_{xy}/I_{xx}$) due to OHE is expressed as $R_{OHE} = \mu_0 R_0 H$, where R_0 is the ordinary Hall coefficient.

3.2 Anomalous Hall effect (AHE)

In conducting magnetic materials with uniform magnetization (M), like ferromagnets, an anomalous contribution to the Hall signal is often observed in addition to the OHE, which is called "anomalous" Hall effect (AHE). AHE occurs intrinsically due to the spin-orbital coupling as a result of the "fictitious" magnetic field added by the magnetization and thus the anomalous Hall resistance is proportional to the magnetization M_z, expressed as:

$$R_{AHE} = R_A M_z \tag{4}$$

where R_A is the anomalous Hall coefficient. However, apart from intrinsic source (i.e., M), there can be extrinsic sources (impurity scattering like, side-jump or skew scattering) can contribute to the anomalous Hall signal. Information of the contributing source (intrinsic or extrinsic) is generally contained in the R_A: For intrinsic sources, $R_A \propto R_{xx}^2$; and for extrinsic sources $R_A \propto R_{xx}$ (where R_{xx} is the longitudinal resistance of the sample).

Thus AHE basically occurs due to the effect of magnetic field on the spin orientation in momentum (reciprocal) space and not in position (real) space that makes it different from topological Hall effect which is associated with spin arrangements in real space (**Figure 4**).

3.3 Topological Hall effect (THE)

In materials with topologically protected real space chiral spin textures, such as skyrmions, strong exchange coupling occurs between the moments of the conduction electrons and the local magnetic moments at each sites. The spin of the electrons

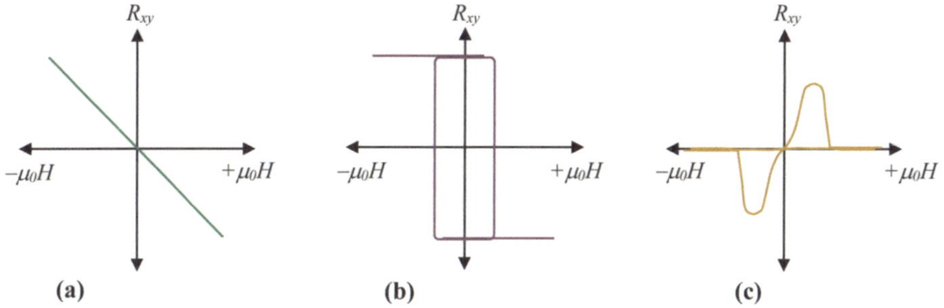

Figure 4.
Schematics of THE typical signatures of Hall resistances arising from different types of Hall effects: (a) OHE only, (b) AHE only, and (c) THE only.

adiabatically follow the spin texture and thus pick up a quantum-mechanical Berry phase, which obviously is sensitive to the topology of the texture. The Berry phase is considered to give rise to another "virtual" (or emergent, or effective) spatially varying magnetic field, b_{eff}. As a result, the conduction electrons "feel" the emergent virtual magnetic field (b_{eff}) arising from the spin textures and are deflected perpendicularly to the applied current direction, resulting in the so-called topological Hall effect (THE). This THE appears as a bump or dip during the hysteretic Hall resistivity measurements. Thus, unlike OHE or AHE, it is obviously neither proportional to external magnetic field ($\mu_0 H$) nor to the magnetization (M) of the sample.

In contrast, the topological Hall voltage signal is proportional to emergent magnetic field (b_{eff}). In case of skyrmions, the b_{eff} contains the information of the skyrmion density; it is proportional to n_{sk}, and hence for skyrmions, the THE voltage is inversely proportional to the size of the skyrmions. The topological Hall resistivity can be given by the following equation [11]:

$$\rho_{THE} = \frac{PR_0 b_{eff}}{en} = \frac{P\Phi_0}{d_{sk}^2 en} \tag{5}$$

Here, P- spin polarization of carriers, d_{sk}- distance between skyrmions, n- carrier density and Φ_0 is the quantum flux (h/e). The skyrmion density (n_{sk}) gives rise to the emergent magnetic field as $b_{eff} = n_{sk}\Phi_0$. Thus from Eq. (5), one can say that separation of skyrmions (d_{sk}) varies as $n_{sk}^{-1/2}$. Eq. (5) can thus also be re-written as

$$\rho_{THE} = PR_0 n_{sk}\Phi_0 \tag{6}$$

It must be mentioned here, that, in case of skyrmions, the emergent magnetic field (b_{eff}) is expected to be independent of temperature and often is very high (may range from few tens of Tesla to few hundreds of Tesla) [12].

It is highlighted here that the appearance of distinctive bumps or dips in Hall resistivity signal of such chiral magnetic systems is a direct consequence of stabilization of non-collinear topological spin textures. However, it should be noted that complications in AHE, *e.g.*, arising from the electronic band structure or sign change of the dominant scattering mechanism might influence the Hall resistivity signal equivalently [13], resulting in similar transport features. The emergence of a bump/dip characteristic in the Hall effect, for example, may not hold strict for skyrmions; and thus cannot always suffice as an unambiguous evidence for identifying non-collinear "topological" spin textures [14, 15]. Hence, along with THE detection, other imaging/x-ray based techniques (as mentioned above) might be required to support the claim of skyrmions as the origin of THE.

In any case, the acquired total Hall resistance signal contains the three components: OHE, AHE and THE. It is thus necessary to discuss in detail how to extract individual components in order to eliminate any experimental artifacts [16].

3.3.1 THE signal extraction technique

Step-1: During experiments with unpatterned thin film samples (say, as in **Figure 2(a)**), electrical four-probe (or van der Pauw) contacts are often made with conductive paints or solders by eye estimation, which adds the longitudinal magneto-resistance (MR) component in the Hall resistance due to the inevitable misalignment of the electrodes. This has to be removed, first, by using $R_{xy}(H)\uparrow$ = $[R_{xy}^{raw}(H)\uparrow$ - $R_{xy}^{raw}(-H)\downarrow]/2$, where $R_{xy}^{raw}(H)\uparrow$ stands for the measured raw data in the 4th and 1st quadrants with the magnetic field scanning from -$|\mu_0 H_{max}|$ to +$|\mu_0 H_{max}|$ and $R_{xy}^{raw}(-H)\downarrow$ the measured raw data in the 2nd and 3rd quadrant for the reversed field scanning. Similarly, the $R_{xy}(H)\downarrow$ branch was extracted using $[R_{xy}^{raw}(H)\downarrow$ - $R_{xy}^{raw}(-H)\uparrow]/2$. In this way one can eliminate the longitudinal (MR) contribution due to misalignment of electrodes. Apart from this, another way to eliminate it from appearing in the Hall data is by pattering the thin film samples in the form of Hall bar. This leaves us with pure Hall signal (R_{xy}) which contains the contributions: R_{xy}^{total} = $\mu_0 R_0 H + R_A M_z + P R_0 b_{eff}$.

Step-2: The ordinary Hall resistance component (which varies linearly with the applied magnetic field) is subtracted from the total Hall resistance loop by slope-deduction method at the high field region. This leaves us with the AHE and THE components: R_{xy}^{total} - $R_{OHE} = R_{AHE} + R_{THE}$.

Step-3: Finally the AHE component (R_{AHE}) is subtracted from the above equation to obtain only R_{THE} contribution. There are two methods to find R_{AHE}: one is by measuring out-of-plane magnetization (M_z) and the R_A from longitudinal resistivity (as mentioned in Section 3.2). The other method is by fitting with trial *tanh* or Langevin functions to estimate the AHE, which though is a crude method. Finally what remains is the R_{THE} signal only. The complete process is shown schematically in **Figure 5**. In the next sections we discuss the origin of and ways to manipulate the topological Hall effect in complex oxide heterostructures.

4. Oxide Heterostructures

The quest for magnetic textures in oxides have taken new thrust very recently after the celebrated discovery of chiral spin texture in $La_{1.37}Sr_{1.63}Mn_2O_7$ compound [17]. In this manganite compound, promisingly low threshold current density ($J_c \sim 10^7$ A/m^2) was recorded for their electrically driven motion. Since then, over the years, various layer-by-layer or step-flow oxide growth techniques have emerged, which have enabled the creation of high-quality interfaces for novel heterostructure design. Interface engineering in complex oxide heterostructures has developed into a flourishing field as various intriguing physical phenomena can be demonstrated which

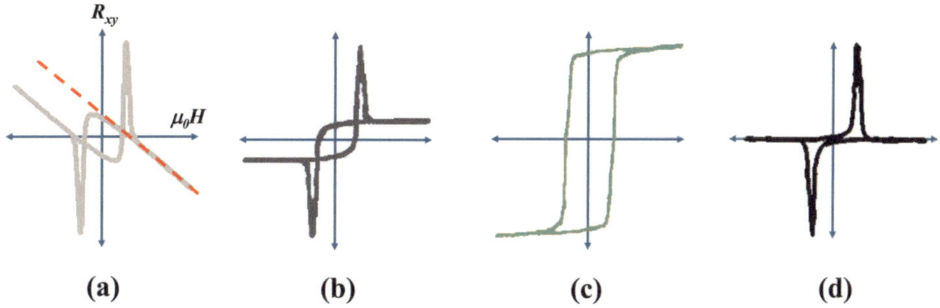

Figure 5.
Steps for extracting THE THE signal from typical Hall data:- (a) gray curve: R_{xy}^{total}. Find OHE by slope-deduction method (red curve) at high field region. (b) after performing Step-2 (mentioned in text), dark gray curve: R_{AHE} + R_{THE}. (c) R_{AHE} determined from magnetization ($M - H$) data. (d) Finally after performing Step-3 (as in text), what remains is THE pure THE resistivity signal (black curve).

are otherwise absent in their constituent bulk compounds [18]. Such capability has opened doors for stabilizing Néel-type skyrmions by engineering the interfacial DMI.

4.1 Growth of oxide heterostructures

Practical applications demand high-quality thin film growth. Owing to some significant advances in the fabrication techniques of high quality oxide heterostructures and the structural compatibility of these kind of complex oxides, the minute tuning of interfacial properties can be routinely achieved. Advanced techniques such as sputtering, pulsed laser deposition (PLD), and molecular beam epitaxy (MBE) have been successfully used to produce high-quality crystalline manganite thin films. Moreover, the presence of a substrate adds another degree of freedom, the external pressure in the form of stress. This is why one of the major characteristics of epitaxial thin films is the strain, which induces modifications in the physical properties (structure, transport and magnetic order) with respect to the bulk.

On the other hand, most of the interfacial emergent properties are highly sensitive to the structure of the interfaces/surfaces, since they arise out of various modified exchange interactions across the interfaces or at the surfaces/terminations (as we will soon explore in upcoming sections of this chapter), and may easily get lost due to presence of slightest of defects.

Intuitively, one has to create a very clean engineered surface or interface to play with such emergent phenomena.

5. Interfacial emergent properties of oxide Heterostructure

As discussed in earlier sections that THE arising from stabilized chiral spin structures or skyrmions require DMI. Considering that the DMI arises from spin-orbit coupling combined with broken inversion symmetry, it is possible to artificially introduce DMI at the surface/interface of complex oxide heterostructures, as discussed below.

Although the individual layers as whole may not show exotic/non-trivial spin structures, various mechanisms like strain, exchange interaction, proximity effect, *etc.*, at the interfaces or surfaces may induce such phenomena and hence, in these heterostructures these properties are called "interfacial emergent" properties.

5.1 Origin and tuning of interfacial DMI

Skyrmions were observed in epitaxial ultrathin ferromagnetic films in the proximity of heavy metal layers, which are subject to giant "emergent" DMIs induced at the interface that breaks inversion symmetry and the strong SOC with neighboring heavy metal. The first investigated system in this class showing "emergent" DMI was heterostructure of Fe monolayer grown on Ir(111) [19].

It is well know that 3*d* transition metal ions rarely show any SOC effect and thus are unlikely to show DMI alone. Rather 4*d* and 5*d* ions are a very good candidate for DMI as they bear inherent strong SOC. For example, $SrRuO_3$ (SRO) (a 4*d* transition metal oxide) has a SOC strength of ~150 m*e*V, while 5*d* transition metal oxide $SrIrO_3$ (SIO) has SOC strength of ~450 m*e*V, 3 times larger than that of SRO [20].

Thus 4*d* and 5*d* based oxides provide platform for engineering the interfacial DMI along with the manganites (Mn based (3*d*) oxides), which have myriad of magnetic phases. The most versatile candidates are $SrRuO_3$ (SRO), $SrIrO_3$ (SIO), $BaTiO_3$ (BTO) and $La_{2/3}Sr_{1/3}MnO_3$ (LSMO); additionally, they posses very common crystal structure- the ABO_3 Perovskite structure and hence high quality layer-by-layer growth can be easily achieved. In the following sub-sections, some recent progress in this field is presented, highlighting the origin and tuning mechanisms of the interfacial DMI, detected electrically through topological Hall effect.

5.1.1 Bilayer structures with strong SOC based oxide

The first investigated system in this class was epitaxial bilayer heterostructure of SRO and SIO in 2016 [11]. SIO is a paramagnet which can host 5*d* electrons with strong spin-orbit coupling [21], and SRO is an itinerant ferromagnet with a Curie temperature (T_C) of ~160 K [22]. The bilayers were grown by PLD on $SrTiO_3$ (STO) substrate with different SRO layer thickness (m, basically the number of unit cells) ranging from 4 to 7 unit cells, while keeping the SIO layer thickness fixed at 2 unit cells (uc). So in the layered structure STO//SRO (m uc)/SIO (2 uc), the interface between SRO and SIO provided the broken inversion symmetry and the SIO layer provided the strong SOC, mimicking an effective "heavy metal" layer. The systematics of the magnetic and transport properties of the bilayers were governed by SRO and could be precisely controlled by tuning SRO layer thickness, m.

Using techniques mentioned as in Section 3.3.1 above, THE component was extracted from the Hall resistivity data for each measurements, where AHE components were derived by measuring the Kerr rotation angle (the Kerr signal magnitude is proportional to M and thus to AHE as well). Among the samples, the THE was highest for m = 4 uc. With increasing m to 5 uc, similar peak structure related to THE could be observed, however, the THE component decreased. Upon further increasing m, THE decreased and vanished at only m=7 uc.

At the high magnetic field regions (say ±9 T) where the magnetization gets saturated, the spins at the Ru sites align ferromagnetically (in corresponding directions depending on field), which lead to absence of spin chirality and the Hall resistivity in those high field regions could be attributed to AHE only. Decreasing the field from 9 T to 0 T, it remains in the ferromagnetic (FM) state with positive magnetization, which corresponds to the observed finite AHE and the negligible THE. With further decrease of the magnetic field from 0 to -0.8 T, absolute value of THE sharply increases upto -0.06 T and then gradually decreases to zero at around -0.4 T. This field (-0.4 T) coincides with the field at which the hysteresis in M vs. $\mu_0 H$ loop closes. This is indicative of the fact that some specific spin structure with finite scaler spin chirality might have been induced when the FM spins started to reverse. The appearance of hysteresis in magnetization and THE in Hall resistivity at simultaneous field range indicates a coexistence between co-planar FM phase and the scalar spin chiral phase.

As discussed earlier, the most plausible chiral spin texture responsible for THE is magnetic skyrmion which gives rise to the emergent magnetic field. This emergent magnetic field strongly affects the electron transport and marginally affects the magnetization.

Due to the broken inversion symmetry at SRO/SIO interface and the strong SOC of SIO, the finite DM vector pointed in the in-plane direction, which might give rise to a Néel-type magnetic skyrmion. The fact, THE appeared only when $4 \geq m \leq 6$, suggests that it was derived from interfacial DMI.

It is worth mentioning that in contrast to the very narrow $T - H$ window of THE in bulk B_{20} compounds (which exhibit Bloch type skyrmions), the THE in oxide bilayer heterostructures was found within a wide T range upto around 90 K. The thickness variation also revealed the stabilization of 2-D nature of the Néel-type skyrmions.

Moreover, utilizing Eqs. (5) and (6) with P $\sim$ -9.5 % for SRO [23], the distance between skyrmions in these bilayers was approximately estimated to be around 10 nm to 20 nm. This provided an estimation of the length scales of the skyrmions, which was always larger than the film thickness ($\sim$2 nm in this case) and confirmed the two-dimensional nature of the skyrmions.

5.1.2 Single layer ultrathin film

In another contrasting work, a few years ago, THE was observed in a single layer SRO film grown on STO insulating substrate; but this time, astonishingly, without the presence of any 5*d* based metal oxides like $SrIrO_3$ [24]. Here the thickness of SRO films ranged from 3 nm to 10 nm, all having tetragonal structure. From structural characterizations it was observed that the strained SRO films in tetragonal phase showed rotation of the RuO_2 octahedra about the c-axis, which could persist upto several tens of nm.

In this case also, the THE appeared in the vicinity of the magnetization reversal, which was consistent with the previous result as mentioned above. Furthermore, the amplitude of the THE resistivity increased with increasing T from 20 K to 80 K, and above 85 K, THE signal vanished. The itinerant magnetic property and the strong perpendicular magnetic anisotropy in these single layer SRO films are indicative of the fact that the non-trivial topological spin texture responsible for the THE must be the Néel skyrmion.

Needless to say, the terminating surface provides the broken symmetry for this single layer system, while Ru ions at the surfaces are the sources of SOC. This gave rise to the DMI, where DM vector also point along the film plane. The oxygen octahedral rotation due to combination of substrate induced strain and natural termination of top layer has a significant effect on THE of the SRO single layer.

Further, in order to elucidate on the origin of SOC of Ru ions at the surface layers, Hall transport was measured in presence of extra electric field, provided by ionic liquid gating. Although the electric field generated from the ionic liquid could be much higher as compared to that with conventional voltage gating, the penetration depth of the electric field might only be a nm or less due to the high carrier density of the SRO. It was observed that upon applying negative gate bias, the THE diminished. The reason is as follows: Upon applying gate bias, the momentum space around the Fermi level of SRO changed [25]. The electric potential gradient near the SRO surface resulted in a change in the inversion asymmetry, which modulated the SOC in that region. This is reminiscent of the phenomenon observed in case of Rashba-type band splitting and spin splitting [26, 27]. The combined contribution of change in SOC and change in inversion asymmetry affected the DMI, which resulted in the enhancement or suppression of THE with the polarity of gate bias.

As earlier, in the single layer SRO films, the emergence of the THE was observed at reduced dimensions, for thickness 3 nm to 6 nm but not in 10 nm film. This, along with the gate voltage modulation of the THE indicated that 2-D Néel skyrmions formed at the surface of the SRO single layer films.

5.1.3 Bilayer heterostructure with ferroelectric layer

Now that we have explored the idea of tuning the THE by applying electric field through ionic gating on SRO single layer films, one might be intuitively tempted to utilize ferroelectric materials in the proximity of SRO layers, as ferroelectric materials could be easily manipulated by electric field since they have the added advantage of inherent inversion symmetry breaking.

Heretofore, tuning of interfacial DMI due to lattice distortions driven by ferroelectricity at the $SrRuO_3/BaTiO_3$ (BTO) interface was investigated in detail [28]. Ultrathin bilayer heterostructures of SRO/BTO were prepared on STO substrate with SRO as the bottom layer. The SRO layer thickness t_{SRO} was varied between 4 and 8 uc and BTO layer thickness t_{BTO} was varied between 3 and 20 uc.

FE-driven ionic displacements in BTO could cross the interface and continue for several unit cells into SRO, a phenomenon what is known as the FE proximity effect [29]. The inevitable lattice distortion due to FE proximity effect could break the inversion symmetry of the SRO structure near SRO/BTO interface. The degree of this inversion symmetry breaking in SRO could be elucidated as the vertical ionic displacement between Ru and O (δ_{Ru-O}) ions in the RuO_2 plane, taking into account a c-axis-oriented FE polarization. In such a case also, one can expect a emergent DMI in the ferroelectrically distorted SRO lattice, where DM vector should lie in-plane, perpendicular to the Ru-O-Ru chains. Accordingly, it is expected that this in-plane DMI at the vicinity of SRO/BTO interface might stabilize Néel-type magnetic skyrmions, giving rise to emergent THE.

As obvious, The Hall signals of SRO/BTO samples depend strongly on the individual layer thicknesses. In this case, the THE signal persisted upto $T \sim 80$ K, but for a much larger field range, with peak around ± 1.65 T and vanished at around ± 3.9 T.

To estimate the basic skyrmionic properties from the THE signal, the evolution of n_{sk} (estimated from Eq. (6)) with sample structures (basically t_{SRO} and t_{BTO}). With increasing t_{SRO} from 4 to 6 uc, n_{sk} decreased rapidly by almost an order of magnitude, which implied that the DMI was stronger in the vicinity of the SRO/BTO interface. On the other hand, n_{sk} also decreased with decreasing t_{BTO} below 8 uc due to suppression of the FE polarization. This trend further demonstrated that skyrmions could be driven by FE, through the proximity effect.

Further, because of a strong correlation between the interfacial DMI and FE, the manipulation of skyrmions with changes in ferroelectric polarization of BTO layer was also explored. For this, the BTO layer had to be electrically poled into pre-designed domain structures which led to local switching of the BTO polarization in the patterned Hall bars of the bilayer structures. This was performed using a conducting AFM tip at room temperature; voltage bias of -(+)8 V led to upward (downward) poling; and then the Hall measurements were performed at low temperatures. The local switching of the BTO polarization (in the out-of-plane/c-axis direction in this case) due to poling led to changes in the δ_{Sr-Ru} near the SRO/BTO interface. This further led to changes in δ_{Ru-O} from the pristine state, which remarkably affected the THE signal. It was observed that the uniformly upward-poled Hall bar exhibited enhancement in polarization which slightly enhanced the ρ_{THE} as compared with that in pristine FE state. However, upon complete downward poling, polarization changed and as a result ρ_{THE} decreased by about 80 %. The magnetic field and temperature range of THE also got reduced due to downward poling which signified a substantial reduction in n_{sk}.

Thus, by downscaling the FE domain size in the above procedures, one might not only be able to tune the overall skyrmion properties microscopically but also could control the nucleation/deletion of individual skyrmions.

5.1.4 Bilayers heterostructures composed of manganites

Enlightened by the observations thin films and heterostructures as mentioned in previous sub-sections, an intuitive question appears: whether the emergent THE observed in these complex oxide heterostructures only appear specifically in the presence of SRO, which itself exhibits a relatively large SOC with 4*d* transition-metal Ru? Hence, it became desirable to investigate the magnetotransport properties in heterostructures composed of SIO and other magnetic oxides. In this subsection, we discuss briefly the emergent THE observed for the first time in the 3*d* perovskite $La_{0.7}Sr_{0.3}MnO_3$ (LSMO) in SIO/LSMO heterostructures deposited epitaxially on the (001) STO substrates [30].

Although the magnetic easy axis of LSMO is commonly known to lie in-plane along < 110 > direction on STO (001)-oriented substrate, it was reported that the easy axis could shift to < 100 > direction when interfaced with a SIO layer of thickness < 5 uc due to strong SOC of SIO layer [31]. Hence the heterostructures used for the study of THE were composed of 2 uc SIO followed by 6–10 uc LSMO; the final structure being STO//SIO (2 uc)/LSMO (*m* uc), *m* = 6,8,10.

As shown in **Figure 6**, the emergent THE peak appeared in a wide temperature range of upto 200 K, and exhibited a gradual broadening with decreasing temperature, very similar trend seen in other oxide heterostructure systems as mentioned above and in typical skyrmion hosting materials like the B_{20} alloys [7]. Moreover, the giant THE resistivity of $\sim 1.0\ \mu\Omega$.cm (in average) was significantly higher than those reported in complex oxide heterostructures composed of SRO/SIO, SRO/BTO or SRO films, demonstrating the feasibility of using the proximity effect of SIO to create novel spin textures in oxide magnetic heterostructures.

To confirm the interfacial origin of the THE observed in the LSMO/SIO heterostructures, Hall effect measurements were performed on separate LSMO single layer films and on LSMO (8 uc)/STO (2 uc)/SIO (2 uc) trilayer heterostructures, both

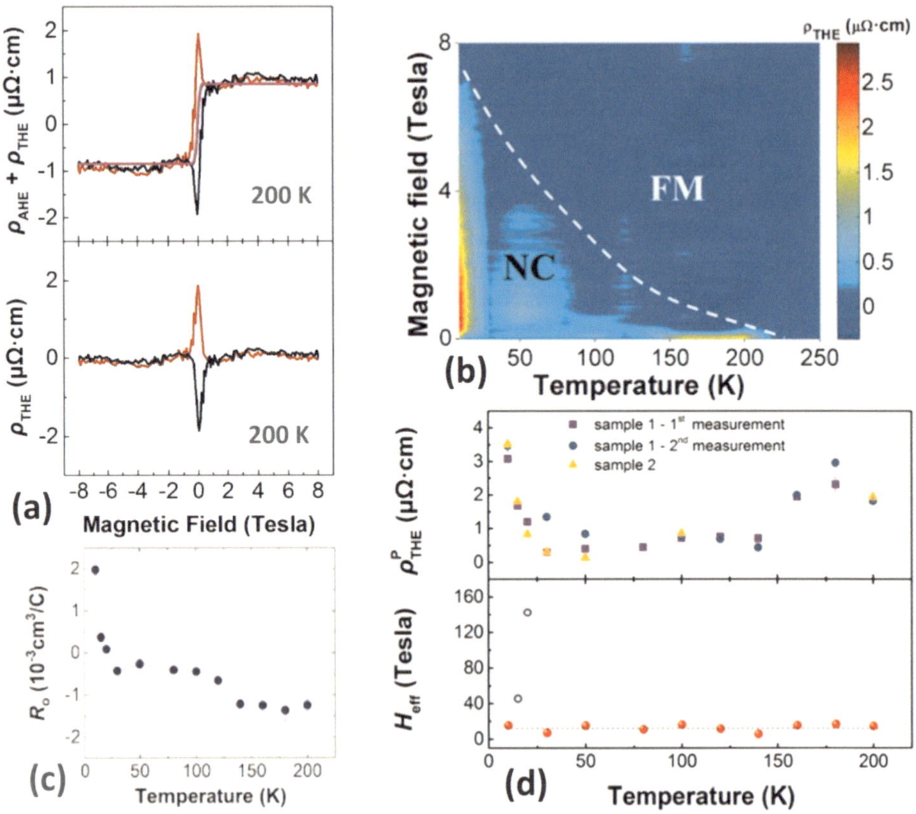

Figure 6.
(a) Upper panel: AHE + THE combined signal at 200 K for SIO (2 uc)/LSMO (8 uc) heterostructure; lower panel: Only THE resistivity at 200 K for THE same sample. (b) H-T phase diagram of THE THE resistivity for same heterostructure generated from Hall measurements at various temperatures. (c) Ordinary Hall co-efficient at various temperatures obtained from Hall measurements for THE same heterostructure. (d) Upper panel: Peak value of THE THE resistivity at various temperatures measured several times and on several samples of THE same heterostructure SIO (2 uc)/LSMO (8 uc), that showed THE reproducibility of THE THE signal; lower panel: The effective fictitious magnetic field obtained from Eq. (5), which was almost constant with temperature, confirmed that the emergent THE was due to formation of skyrmions as a result of interfacial DMI. [reprinted (adapted) with permission from [30]. Copyright (2019), American Chemical Society.]

grown on STO substrates. Distinctively, no THE signals were observed in those two samples (single layer and trilayer). Although the absence of THE in LSMO single layer films was upto expectation, the absence of THE in the trilayer samples could only be ascribed to the inserted non-magnetic insulating STO layer, which interrupted the strong interfacial DMI between the LSMO and SIO layers.

The effective emergent magnetic field b_{eff}, associated with the real space Berry phase, was also found to be independent of temperature upto 200 K as shown in **Figure 6(d)**, with an average value of around 12 Tesla. Such a strong stability against temperature indicated that the origin of emergent THE might be due to skyrmions.

In another work, interfacial atomic layer control of THE by deliberately controlling the competition between chiral DMI and intrinsic collinear FM in 3*d*-5*d* heterostructures composed of $LaMnO_3/SrIrO_3$ was demonstrated [32]. This interfacial symmetry control led to a large THE, which was believed to be originated from a highly robust chiral magnetic phase, potentially hosting skyrmions.

6. Further scope

Heterostructure engineering in complex oxide systems with polycrystalline heavy metal layers like, Pt or W instead of strong SOC based oxides is now an active sub-domain of research which is worth investigating, since the interfacial DMI is key for stabilization of Néel type skyrmions.

It has long been felt for the utilization of oxides in the field of flexible electronics. This requirement has triggered the research on growth and characterization of complex oxide thin films like on various flexible substrates like, mica, polymide tapes, *etc.* [33, 34] using common epitaxial lift-off techniques with the aid of sacrificial layers. A thorough and exhaustive study on the film growth and transfer on flexible substrates, their characterization (using AFM or electron microscopy techniques) and further establishing a direct relationship between strain and magnetic, magnetotransport properties are now being investigated in details.

7. Conclusions

To conclude, we presented the detection technique of interfacial emergent phenomena like interfacial DMI by means of topological Hall effect (THE) in complex oxide heterostructures, which only provides a tell-tale sign of the existence of chiral textures.

We also presented recent progresses towards various methods of achieving and tuning the inversion asymmetry and spin-orbit coupling to tailor minutely the interfacial DMI in those kind of thin film heterostructures, based on some recent works. To illustrate, we presented examples from each of the following methods: tuning the structure through atomic (unit cell) layer control at interface of *4d*/*5d* based heterostructures (SRO/SIO); harnessing octahedral rotation due to strain in ultrathin single layer SRO films; utilizing ferroelectric polarization to tune the RuO_2 octahedra in SRO/BTO heterostrcutures; using the exchange interactions among the Mn spins in (*3d*-*5d*) based heterostrcutures (LSMO/SIO) and superlattices (LMO/SIO) by means of unit cell modification along with the strong SOC of the *5d* layers at the interfaces.

For the device application thin film heterostructures are very important. We expect that the recent progresses will aid to the future skyrmion based devices for the manipulation by current, electric field and/or by some other techniques as well, so that the information can be used for the memory devices and logic implementation.

Acknowledgements

The author acknowledges Dr. Samik DuttaGupta and Mr. Arnab Bhattacharya of SINP, Kolkata, for scientific discussions. Finally, the author acknowledges TCG CREST, Kolkata, for the Post-Doctoral fellowship.

Conflict of interest

The author declares no conflict of interest.

Author details

Snehal Mandal
Centre for Quantum Engineering, Research and Education (CQuERE), TCG Centres for Research and Education in Science and Technology (TCG-CREST), Kolkata, India

*Address all correspondence to: snhlmandal@gmail.com

References

[1] Dzyaloshinsky I. A thermodynamic theory of "weak" ferromagnetism of antiferromagnetics. Journal of Physics and Chemistry of Solids. 1958;**4**(4): 241-255. DOI: 10.1016/0022-3697(58)90076-3

[2] Moriya T. Anisotropic superexchange interaction and weak ferromagnetism. Physical Review. 1960;**120**(1):91. DOI: 10.1103/PhysRev.120.91

[3] Liu Z, Burigu A, Zhang Y, Jafri HM, Ma X, Liu E, et al. Giant topological Hall effect in tetragonal Heusler alloy Mn_2PtSn. Scripta Materialia. 2018;**143**: 122-125. DOI: 10.1016/j.scriptamat.2017.09.024

[4] Mühlbauer S, Binz B, Jonietz F, Pfleiderer C, Rosch A, Neubauer A, et al. Skyrmion lattice in a chiral magnet. Science. 2009;**323**(5916):915-919. DOI: 10.1126/science.1166767

[5] Yu X, Kanazawa N, Onose Y, Kimoto K, Zhang W, Ishiwata S, et al. Near room-temperature formation of a skyrmion crystal in thin-films of the helimagnet FeGe. Nature Materials. 2011;**10**(2):106-109. DOI: 10.1038/nmat2916

[6] Meng K-Y. Magnetic Skyrmions in Oxide Thin Film Heterostructures [Thesis]. USA: The Ohio State University; 2019

[7] Neubauer A, Pfleiderer C, Binz B, Rosch A, Ritz R, Niklowitz P, et al. Topological Hall effect in the *A* phase of MnSi. Physical Review Letters. 2009; **102**(18):186602. DOI: 10.1103/PhysRevLett.102.186602

[8] Soumyanarayanan A, Raju M, Gonzalez Oyarce A, Tan AK, Im M-Y, Petrović AP, et al. Tunable room-temperature magnetic skyrmions in Ir/Fe/Co/Pt multilayers. Nature Materials. 2017;**16**(9):898-904. DOI: 10.1038/nmat4934

[9] Hall EH et al. On a new action of the magnet on electric currents. American Journal of Mathematics. 1879;**2**(3): 287-292. DOI: 10.2307/2369245

[10] Hall EH, XVIII. On the "rotational coefficient" in nickel and cobalt, the London, Edinburgh, and Dublin philosophical magazine and journal of. Science. 1881;**12**(74):157-172. DOI: 10.1080/14786448108627086

[11] Matsuno J, Ogawa N, Yasuda K, Kagawa F, Koshibae W, Nagaosa N, et al. Interface-driven topological Hall effect in $SrRuO_3 - SrIrO_3$ bilayer. Science Advances. 2016;**2**(7):e1600304. DOI: 10.1126/sciadv.1600304

[12] Jiang W, Chen G, Liu K, Zang J, Te Velthuis SG, Hoffmann A. Skyrmions in magnetic multilayers. Physics Reports. 2017;**704**:1-49. DOI: 10.1016/j.physrep.2017.08.001

[13] Fan W, Ma L, Zhou S. Sign change of skew scattering induced anomalous Hall conductivity in epitaxial NiCo (002) films: Band filling effect. Journal of Physics D: Applied Physics. 2015;**48**(19):195004. DOI: 10.1088/0022-3727/48/19/195004

[14] Chakraverty S, Matsuda T, Wadati H, Okamoto J, Yamasaki Y, Nakao H, et al. Multiple helimagnetic phases and topological Hall effect in epitaxial thin films of pristine and Co-doped $SrFeO_3$. Physical Review B. 2013; **88**(22):220405. DOI: 10.1103/PhysRevB.88.220405

[15] Kimbell G, Kim C, Wu W, Cuoco M, Robinson JW. Challenges in identifying

chiral spin textures via the topological Hall effect. Communications Materials. 2022;**3**(1):19. DOI: 10.1038/s43246-022-00238-2

[16] Wang H, Dai Y, Chow GM, Chen J. Topological Hall transport: Materials, mechanisms and potential applications. Progress in Materials Science. 2022;**103**: 100971. DOI: 10.1016/j.pmatsci.2022.100971

[17] Yu X, Tokunaga Y, Kaneko Y, Zhang W, Kimoto K, Matsui Y, et al. Biskyrmion states and their current-driven motion in a layered manganite. Nature Communications. 2014;**5**(1): 3198. DOI: 10.1038/ncomms4198

[18] Lim ZS, Jani H, Venkatesan T, Ariando A. Skyrmionics in correlated oxides. MRS Bulletin. 2021;**46**(11): 1053-1062. DOI: 10.1557/s43577-021-00227-9

[19] Heinze S, Von Bergmann K, Menzel M, Brede J, Kubetzka A, Wiesendanger R, et al. Spontaneous atomic-scale magnetic skyrmion lattice in two dimensions. Nature Physics. 2011; 7(9):713-718. DOI: 10.1038/nphys2045

[20] Mattheiss L. Electronic structure of RuO_2, OsO_2, and IrO_2. Physical Review B. 1976;**13**(6):2433. DOI: 10.1103/PhysRevB.13.2433

[21] Matsuno J, Ihara K, Yamamura S, Wadati H, Ishii K, Shankar VV, et al. Engineering a spin-orbital magnetic insulator by tailoring superlattices. Physical Review Letters. 2015;**114**(24): 247209. DOI: 10.1103/PhysRevLett.114.247209

[22] Schultz M, Levy S, Reiner JW, Klein L. Magnetic and transport properties of epitaxial films of $SrRuO_3$ in the ultrathin limit. Physical Review B. 2009;**79**(12):125444. DOI: 10.1103/PhysRevB.79.125444

[23] Worledge D, Geballe T. Negative spin-polarization of $SrRuO_3$. Physical Review Letters. 2000;**85**(24):5182. DOI: 10.1103/PhysRevLett.85.5182

[24] Qin Q, Liu L, Lin W, Shu X, Xie Q, Lim Z, et al. Emergence of topological Hall effect in a $SrRuO_3$ single layer. Advanced Materials. 2019;**31**(8): 1807008. DOI: 10.1002/adma.201807008

[25] Fang Z, Nagaosa N, Takahashi KS, Asamitsu A, Mathieu R, Ogasawara T, et al. The anomalous Hall effect and magnetic monopoles in momentum space. Science. 2003;**302**(5642):92-95. DOI: 10.1126/science.1089408

[26] Datta S, Das B. Electronic analog of the electro-optic modulator. Applied Physics Letters. 1990;**56**(7):665-667. DOI: 10.1063/1.102730

[27] Manchon A, Koo HC, Nitta J, Frolov SM, Duine RA. New perspectives for Rashba spin–orbit coupling. Nature Materials. 2015;**14**(9):871-882

[28] Wang L, Feng Q, Kim Y, Kim R, Lee KH, Pollard SD, et al. Ferroelectrically tunable magnetic skyrmions in ultrathin oxide heterostructures. Nature Materials. 2018; **17**(12):1087-1094. DOI: 10.1038/s41563-018-0204-4

[29] Guo H, Wang Z, Dong S, Ghosh S, Saghayezhian M, Chen L, et al. Interface-induced multiferroism by design in complex oxide superlattices. PNAS. 2017;**114**(26):E5062-E5069. DOI: 10.1073/pnas.1706814114

[30] Li Y, Zhang L, Zhang Q, Li C, Yang T, Deng Y, et al. Emergent topological Hall effect in $La_{0.7}Sr_{0.3}MnO_3$/

$SrIrO_3$ heterostructures. ACS Applied Materials & Interfaces. 2019;**11**(23): 21268-21274. DOI: 10.1021/acsami.9b05562

[31] Yi D, Liu J, Hsu S-L, Zhang L, Choi Y, Kim J-W, et al. Atomic-scale control of magnetic anisotropy via novel spin-orbit coupling effect in $La_{2/3}Sr_{1/3}MnO_3/SrIrO_3$ superlattices. PNAS. 2016; **113**(23):6397-6402. DOI: 10.1073/pnas.1524689113

[32] Skoropata E, Nichols J, Ok JM, Chopdekar RV, Choi ES, Rastogi A, et al. Interfacial tuning of chiral magnetic interactions for large topological Hall effects in $LaMnO_3/SrIrO_3$ heterostructures. Science Advances. 2020;**6**(27):eaaz3902. DOI: 10.1126/sciadv.aaz3902

[33] Boileau A, Dallocchio M, Baudouin F, David A, Lüders U, Mercey B, et al. Textured manganite films anywhere. ACS Applied Materials & Interfaces. 2019;**11**(40):37302-37312. DOI: 10.1021/acsami.9b12209

[34] Lim ZS, Khoo AKH, Zhou Z, Omar GJ, Yang P, Laskowski R, et al. Oxygen octahedral tilt controlled topological Hall effect in epitaxial and freestanding $SrRuO_3/SrIrO_3$ Heterostructures. arXiv. 2022;**2204**: 06174. DOI: 10.48550/arXiv.2204.06174

Section 3

Electrochemical Applications

Chapter 7

Emerging Thin Films Electrochemical Applications: The Role of Interface

Dongmei Dong

Abstract

Despite the natural cycle of climate change being unavoidable and the reality, history has been telling the living the previous civilizations that have overstressed themselves or pushed the consumption of natural resources to the limit, and the consequence soon shows the alert in the climate. It is a critical period of time to change the current in response to climate change by employing environmentally friendly and emission-free energy technologies. The applications of advanced functional thin films ranging from the quantum level to nano and microscale, from inorganic metal oxides to conductive polymers, have been pushing the rapid development of energy-saving technologies and clean and renewable energy production, storage, and conversion in the past decade. This specific chapter focuses on fundamental- and applied- science on various advanced thin films and their applications in reliable renewable energy devices and/or systems, including but not limited to electrochromics, supercapacitors, fuel cells, flow batteries, electrolysis, triboelectricity, etc. Given that much of the work is realized across interfaces, the spotlight is shielded onto the interface of thin films in electrochemistry with different emerging cutting-edge ongoing research examples.

Keywords: thin solid films, interface, nanomaterials, electrochemistry, electrochromic, fuel cells, renewable energy

1. Introduction

Nowadays, there are full of challenges and of course opportunities to respond to environmental change by replying on zero-emission clean energy technology developments. It is estimated about 40% of the primary global energy consumption lies in buildings, that is. lighting, heating, ventilation, and air conditioning (HVAC) sections [1–4]. The understanding of taking the responsibility for reducing CO_2 emission is one of the main driving forces for the emerging and development of advanced electrochromic (EC) materials and devices today, particularly in terms of the utilization of smart windows or glazing for energy-saving houses, buildings, and skyscrapers [5] in cities. In the past decade, EC energy storage devices have become an emerging type of material in the rapid growing of miniaturized electronic devices, wearables, information encapsulation digital display, etc. [6, 7]. EC-based charging/discharging has gained elevated general public attention recently. Complementary to energy storage, they are able to adjust the relations between humans and sunlight utilizing their tunable and reversible optical response. EC glazing is estimated to reduce energy consumption by about 20% in the building sector [8–15]. Charging capabilities and efficient energy derivation contribute to the energy recycling and have various functions from lighting and LEDs to wearable devices [16]. The visible charge level also provides much convenience to users. Most known renewable energy systems share concurrent electrochemical fundamentals, following thermodynamics but further restricted by kinetics. Enhancing the kinetics limited by nonideal conditions is dependent on the intrinsic understanding at as small a scale as we can.

Even though EC materials are being studied to obtain longer lifetime, the failure fundamentals are still unclear. A succession of chemical, physical, optical, electrical, and mechanical evolution in electrochemical energy devices arising from charge transfer (CT) kinetics has not been yet understood clearly. The potential future applications, with certainty, require a comprehensive deep digestion of the mechanisms beneath. The EC fundamentals and the durability issues should be mastered at a smaller scale. So far, the functional charging materials for small active ions (Li^+)-based cells or systems have been found to be restricted by the instability events occurred at the interface across electrodes and electrolytes [17]. The across-interface activities, durability, and even the physical and nanoscale mechanical variations and related electrical changes remain poorly discovered, imposing a fundamental challenge for electrochemical CT in energy systems. Lots of inorganic transition metal-based oxide complex demonstrates a series of changes covering different aspects, from the physical and chemical, to optical, and even mechanical variations arising from electrochemistry. An EC case, WO_3, for example, changes its appearance from colorless to bright and deep blue colors upon reduction due to Li^+/e^- dual incorporation. In addition to cathodic EC WO_3, other transition metal oxide, such as NiO is placed at the opposite position anode, switching between brownish and light status. In this electrochemical process, the mechanical stress/strain caused by the ion insertion/extraction brings non-neglectable volumetric expansion/shrinkage, which is termed “mechanical breathing.”

In recent years, elevated attention has been paid to renewable energy development, that is. environmentally friendly electrical vehicles for transportation [18]. Proton exchange membrane fuel cells (PEMFCs) demonstrate great potential for developing clean or green transportation technologies, based on oxygen and hydrogen electrochemical reactions by providing a renewable and zero-emission fuel source replacing the traditional combustion engine. However, one of the unaddressed performance and durability issues lies in the Nafion® membrane due to so-called harmful radical attack that brings many electro-mechanical negative effects in fuel cells [19]. These active radicals come from oxygen reduction reactions at overpotentials in this case the oxygen molecules associate with the catalyst, that is. platinum and travel to the polymer layer [20]. Such serious chemical attack is the main reason for polymer chain scission and physical and chemical irreversibility, causing the overall and regional mass loss of the membrane, which will cause producing fluorinated and sulfated species into reactant waste at the outlets. Synchronous fluorinated and sulfated degradation byproducts will accumulate at the exhaust. Because of chemical degradation, the fluoride emission quantity will significantly increase in a short period of time. Especially, the amount of the fluorinated products will significantly at certain accelerated operations, that is. higher temperatures and dry conditions. In one word, the fluoride and sulfate anions emission rates (FER; SER) can be defined as the signature of the membrane loss, and thus the fuel cell degradations status and a sign for predicting the remaining lifetime.

To precisely quantify the polymer breakdown, the fluoride anions of the wastewater coming from the fuel cell membrane are identified as degradation signatures. Detecting fluoride anions and converting the quantity as a sensor signal can be achieved based on specific fluoride-sensitive films. Fabricating highly selective sensors can be obtained by modifying the active sensing region; its integration in the electrical vehicle allows direct knowing ahead of the remaining life of the fuel cell polymers replacing the existing, wasteful, and costly methods of getting the fuel cell engine changed. Unfortunately, up to date, vehicle manufacturers advise users to change the fuel cell engine after covering a certain amount of miles or hours [21]. It may cause waste as potentially functional membranes in good condition can be changed or may be risky; in that case, a membrane in operation has had more degradation than what is theoretically projected according to the driving mileage or the time after experiencing intensive consumptions at extreme destructive real conditions in driving.

Benefiting from their lower cost, smaller size, robustness, and capabilities for continuous real-time monitoring, ion-sensitive field effect transistors have been promising microsensors. Leveraging the fast development of advanced sensor science and technologies, our team has designed and developed microsensors bringing opportunities for incorporating electrode outlets for showing the fuel cell state of health. The excellent leading signature for Nafion®-based fuel cells is the emission of fluoride anions (F^-), coming along with the exhaust streams at the electrodes [22]. The F^- concentration is chosen as a signature, representing the membrane's failure status in a precise way. Monitoring fluoride and converting its quantity into a voltage/current sensor readout is achieved *via* specific fluoride-sensitive nanofilms. It is a good idea to employ fluoride emission rate as a direct simplified but accurate diagnostic method. Specific fluoride-selective films (LaF_3) will be incorporated into a nanoscale layer of films for functionalization. The modification of the active area changes the selectivity/sensitivity of the developed sensors. It is different and complements to current fuel cell test and troubleshooting approaches. At present, some other methods for detecting fluoride exist; however, very few are compatible as a solution for real-time continuous monitoring of the fluoride emission rate (FER) in fuel cell test stands or electrical vehicles. Most of the existing tools are more delicate, costly, limited in detection, and not portable. The commercially available products fluoride ion selective electrode (FISE) is utilized to provide a reference and kind of ground truth to relatively compare the performance of the developed sensors. FISE is capable of detecting fluoride ions at ppb level and requires calibration operation, maintenance, and specific storage procedures, restricting its applications to fuel cells. It is of great significance to dedicate efforts to design and build microsensors for onsite real-time detection of the signature byproducts, and thus fuel cell membrane or polymer loss. The association of the F^- and fuel cell degradation has been established according to the ion quantity measurement to calculate the membrane degradation rate. The ions released from ionomer breakdown are targeted as fluorinated species, termed FER, which is being taken to tell fuel cell membrane failure degree [23].

In complementary to the interfacial studies for electrochemical performance and durability improvements, a unique and fascinating application is to implement the thin films and cantilever platform to envision the activities at the interface at extraordinary spatial and temporal resolution. The research activities regarding this topic are creative and original due to the synergy between the quantitative diagnostic sensor developments and energy-focused fundamental questions related to nanoscale electrochemistry and energy storage, highly demanding to date. When spectroscopy has opportunities to meet nanomechanics, it is hypothesized that the thermal-mechanical cantilever resonator (TMCR) *in situ*/Operando platform will be capable of decoding modern electrochemistry nanoverse based on its exceptional specificity/sensitivity [24]. This multimodal single utility TMCR is conceptually novel and is of great interest to the chemistry and electrochemistry community for nanometer and ultimate atomic scale studies. Its spatial and temporal resolution is up to sub-nm and μs, respectively; the mass sensitivity is six orders of magnitude higher than commercial quartz crystal microbalance (QCM) [25]; illuminating the infrared (IR) light can obtain compositional identification based on the photo-thermal nanomechanical effects, superior to Fourier transform infrared spectroscopy (FTIR) bounded by its opto-output loss [25]. Three focused characteristics of the versatile TMCR are (1) cantilever static bending and resonant dynamics, (2) nanomechanical holography enabled by multifrequency excitation/detection, and (3) IR spectroscopy through photo-thermal nanomechanical response. Uniquely, this single entity enables real-time simultaneous multi-signal output realized by orthogonal vectors. This real-time data acquisition not only surpasses the limitations of current *ex situ* state-of-the-art equipment but also consolidates multiple measurements into a single platform, promoting measurement efficiency and disseminating research accessibility.

2. Interfacial deformation- electrochromic monolithic ultra-thin-film device ITO/WO_3/Ta_2O_5:Li/NiO/ITO

Sustainability development of the society, clean energy demand, performance the cost are driving modern energy science and technologies fast growing. Much progress in these realms is made including electrochromic (EC) materials and devices by improving their performance and durability. Yet, ion insertion/de-insertion induced charge transfer (CT) nanomechanics, that is. repetitive electrode size change and generated stress/strain during electrochemical cycling, termed the "nanomechanical breathing" effect, has remained unexplored. Electro-chemo mechanics is the focus here due to its intimate correlation to the elastic and plastic deformation at the interface. EC transitional metal oxides with intervalence electrons and excellent electrochemical kinetics show dramatic color changes at various valence states upon redox reactions, which enable them as an emerging category of energy storage supercapacitors. WO_3 and NiO films are configurated as the cathode and anode, respectively. The void generation and delamination at the interface account for the device degradation after extended cycles. Nanoindentation mechanical test and electrical kelvin probe force microscopy are employed to investigate the interface. WO_3 with a charge density of ~40mC/cm^2 shows a ~45% increase in the elastic Young's modulus compared to the pristine transparent state. The redox-induced surface stress originating from ion adsorption and CT is revealed. Noteworthy, upon charging (coloration; lithiation) with a fair smooth surface status giving about 3.4 times more electrostatic surface potential, the electrical work function of the films surprisingly becomes lower because of the dominant effect of the dipole layer potential against the chemical potential. From a physics point of view, it is suggested that the interatomic cohesive energy and equilibrium distance

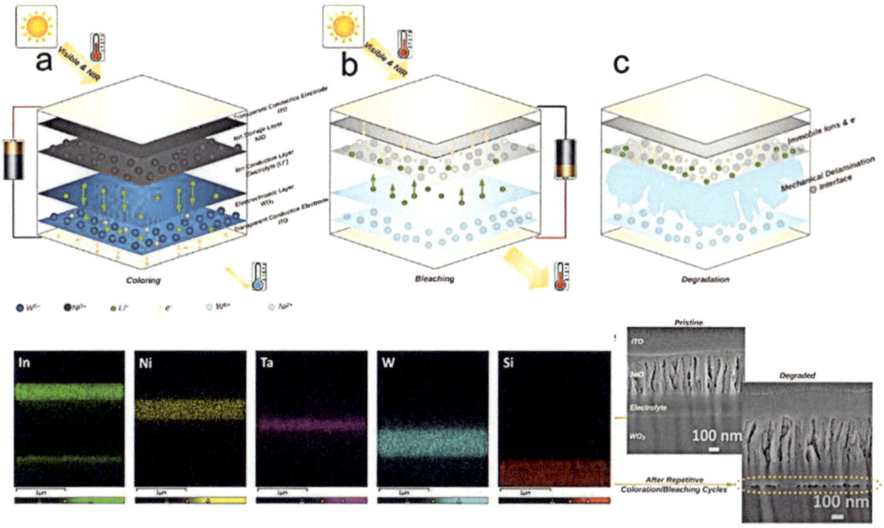

Figure 1.
Author's work on electrochromic film interfacial nanomechanical deformation [26].

increase lead to the mechanical deformation in the long-term charging/discharging. The dependence of surface potential, stress, work function, and cohesive energy on electrochemical kinetics is interpreted. It provides fundamental insights into electro-chemo mechanics and interdisciplinary concerted interfacial effects at the nano/atomic level (**Figure 1**).

3. Interfacial manipulation- sensing purpose: Diagnostics tool for PFSA-fuel cells

Fuel cell electrical vehicles have drawn much attention due to the elevated environmental concern and governmental initiatives on hydrogen infrastructures recently. Perfluorosulfonic acid (PFSA) fuel cells have great potential to be employed for vehicles based on oxygen and hydrogen electrochemical reactions. Unfortunately, one of the limiting factors to the performance is the PFSA membrane. The radical attack and associated irreparable electro-mechanical damage in degradation result in the global and local thinning of the ionomer, causing fluorinated and sulfated degradation chemicals into reactant outlet streams; the options in the market lack specific diagnostics and a legitimate indication of the exact time that the membrane should be changed. To address the need, this work emphasizes developing an onsite sensor for quantifying the degradation by detecting fluoride in effluent water on the basis of surface functionalization and interface manipulation. The sensor's specificity/sensitivity has been achieved in real-time at a sub 10 ppb level, relying on the spin-coating deposition of ~50 nm sensing membrane LaF_3 and post-annealing procedures. The multimodal data collection is obtained, including the characterizations of open circuit potential, cyclic voltammetry, chronoamperometry, and differential pulse voltammetry, demonstrating a consistent linear decrease of Faraday current from the established redox marker, while favoring the capacitive behaviors at the interface by absorbing F- ions (**Figure 2**).

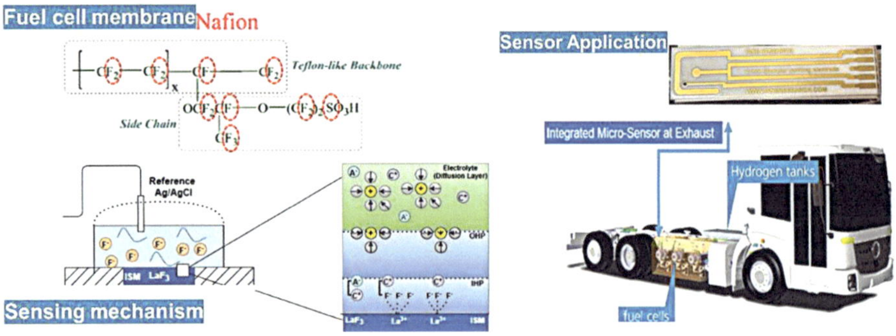

Figure 2.
Author's work on sensor developments based on the capacitive process at the interface [27].

4. Interfacial *in situ*/operando envision- spectroscopy meets nanomechanics: To decode interfacial nanoverse on a thermal-mechanical cantilever resonator (TMCR)

To decode the interface across electrode and electrolyte, an *in situ*/Operando envision tool is highly on-demand. It provides the potential to enable real-time simultaneous multi-signal output based on a conceptually novel single entity, a "thermal-mechanical cantilever resonator (TMCR)." TMCR in situ/Operando envisioned tool is creative and original due to the synergy between the quantitative diagnostic sensor developments and energy-focused fundamental questions related to interface, nanoscale electrochemistry, and energy storage, which is critically demanding to date. It is expected that thin-film-based TMCR will be capable of decoding interfacial charge transfer nanoverse based on exceptional specificity/sensitivity. Three targeted characteristics of the versatile TMCR are (1) cantilever static bending and resonant dynamics, (2) nanomechanical holography by multifrequency excitation/detection, and (3) infrared (IR) spectroscopy *via* photo-thermal nanomechanical response. It is complementary to conventional macroscopic characterizations (**Figure 3**).

Based on its nano/micro dimension, geometric feature, and thermomechanical effects, TMCR provides extraordinary sensitivity and orthogonal vector readouts. The spatial and temporal resolution is up to sub-nm and μs, respectively; the mass sensitivity is six orders of magnitude higher than commercial costly quartz crystal microbalance (QCM); illuminating the infrared (IR) light can obtain compositional identification through photo-thermal nanomechanics, superior to Fourier transform IR spectroscopy (FTIR) bounded by the opto-output loss. The ultrasensitive and informative TMCR is expected to reveal the electrochemical dynamics and unclear fundamental mechanistics at the nanoscale, generally uncaptured with classic macroscopic measurements. It will assist in decoding (i) real-time charge transfer nanomechanics, that is. repetitive electrode size change during electrochemical cycling, termed "nanomechanical breathing," (ii) inhomogeneities, local ion concentration, and other locality properties based on nanomechanical holography, and (iii) quantifying minute changes in metastable intermediates, competing species, and impurities. TMCR real-time data acquisition not only surpasses the limitations of current *ex situ* state-of-the-art equipment but also consolidates multiple

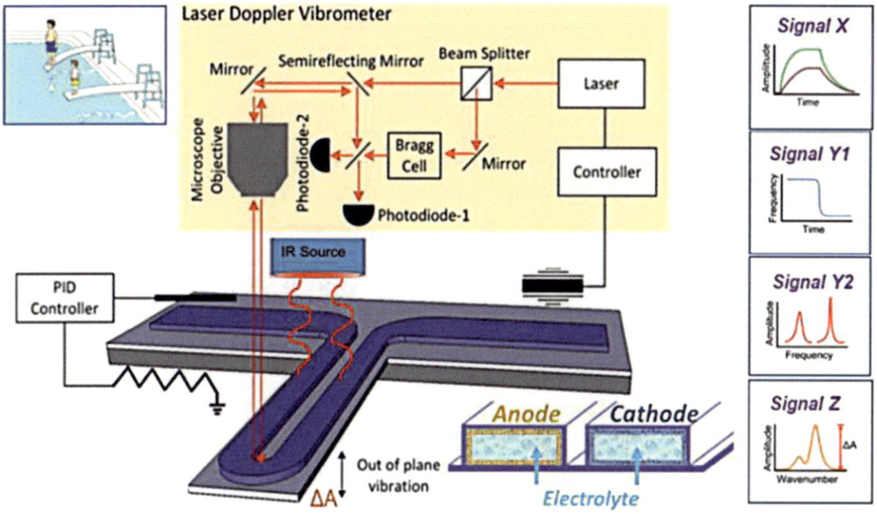

Figure 3.
Thermo-mechanical cantilever resonator (TMCR): Multimodal single entity at micro-dimension as electrode/electrolyte interface envision tool.

measurements into a single platform, promoting measurement efficiency and disseminating research accessibility, not limiting to the thin films and electrochemical energy device research.

5. Conclusions

This chapter focuses on the interface in thin films in electrochemistry and renewable energy applications, which involves fundamental science and challenges from the delamination issue at the interface, and thus the full device degradation, the electrochemical sensing applications based on the interactions and communications across the interface to an advanced envision measurement tool for capturing the minute changes occurred in the electrode/electrolyte. Bridging the gap between the electrochemo nanomechanics and sensing capabilities at the interface with renewable energy applications would require interdisciplinary research with surface physics/chemistry, materials science, nanotechnologies, and engineering. Therefore, the research area can bring scientists and researchers from diverse backgrounds, including chemistry, physics, materials science and engineering, electrical engineering, and mechanical engineering.

Acknowledgements

This work is partially supported by the Million Mile Fuel Cell Truck (M2FCT) consortium, funded by the Department of Energy (DoE), US, under contract No. DE-AC02-05CH11231. The author would like to specially thank Dr. Thomas Thundat for the insightful discussions on thermal-mechanical cantilevers.

Conflict of interest

The authors declare no conflict of interest.

Author details

Dongmei Dong
Physics and Astronomy Department, Rowan University, Glassboro, New Jersey, United States

*Address all correspondence to: dongd@rowan.edu

References

[1] Sathiy M, Prakash AS, Ramesha K, Tarascon JM, Shukla AK. Journal of the American Chemical Society. 2011;**133**:16291

[2] Dinh TM, Achour A, Vizireanu S, Dinescu G, Nistor L, Armstrong K, et al. Nano Energy. 2014;**10**:288

[3] Wei Q, DeBlock RH, Butts DM, Choi C, Dunn B. Energy and Environmental Materials. 2020;**3**:221

[4] Priyadarshini BG, Aich S, Chakraborty M. Thin Solid Films. 2016;**616**:733

[5] Zhang S, Cao S, Zhang T, Fisher A, Lee JY. Energy & Environmental Science. 2018;**11**(10):2884

[6] Mondal S, Ninomiya Y, Higuchi M. ACS Applied Energy Materials. 2020;**3**(11):10653

[7] Mondal S, Santra DC, Ninomiya Y, Yoshida T, Higuchi M. ACS Applied Materials & Interfaces. 2020;**12**(52):58277

[8] Barile CJ, Slotcavage DJ, Hou J, Strand MT, Hernandez TS, McGehee MD. Joule. 2017;**1**:133

[9] Wen RT, Granqvist CG, Niklasson GA. Nature Materials. 2015;**14**:996

[10] Liang X, Guo S, Chen M, Li C, Wang Q, Zou C, et al. Materials Horizons. 2017;**4**:878

[11] Lin H, Yang L, Jiang X, Li G, Zhang T, Yao Q, et al. Energy & Environmental Science. 2017;**10**(6):1476

[12] Yang X, Zhu G, Wang S, Zhang R, Lin L, Wu W, et al. Energy & Environmental Science. 2012;**5**:9462

[13] Yang P, Sun P, Chai Z, Huang L, Cai X, Tan S, et al. Angewandte Chemie (International Ed. in English). 2014;**53**:11935

[14] Xia X, Ku Z, Zhou D, Zhong Y, Zhang Y, Wang Y, et al. Materials Horizons. 2016;**3**:588

[15] Liu L, Du K, He Z, Wang T, Zhong X, Ma T, et al. Nano Energy. 2019;**62**:46

[16] Tsai WY, Wang R, Boyd S, Augustyn V, Balke N. Nano Energy. 2021;**81**:105592

[17] Cannarella J, Leng CZ, Arnold CB. On the coupling between stress and voltage in lithium-ion pouch cells. Energy Harvesting and Storage: Materials, Devices, and Applications V, SPIE. 2014;**9115**:69

[18] Grieshaber D, MacKenzie R, Vörös J, Reimhult E. 2008. 8(3):1400.

[19] Zhang J, Jiang G, Cumberland T, Xu P, Wu Y, Delaat S, et al. InfoMat. 2019;**1**(2):234

[20] IEA U. Global Energy Review. 2020. Available from: https://www.iea.org/countries/ukraine

[21] Spears A, Rockward T, Mukundan R, Garzon FH. ECS Transactions. 2020;**98**(9):407

[22] Spears A, Rockward T, Mukundan R. ECS Transactions. 2019;**92**(8):467

[23] Elgrishi N, Rountree KJ, McCarthy BD, Rountree ES, Eisenhart TT, Dempsey JL. Journal of Chemical Education. 2018;**95**:197

[24] Tetard L, Passian A, Thundat T. New modes for subsurface atomic force microscopy through nanomechanical coupling. Nature Nanotechnology. 2010;**5**(2):105-109

[25] Burg TP, Godin M, Knudsen SM, Shen W, Carlson G, Foster JS, et al. Weighing of biomolecules, single cells and single nanoparticles in fluid. Nature. 2007;**446**(7139):1066-1069

[26] Dong D, Lou L, Lopez KO, Agarwal A, Bhansali S. Revealing nanomechanical deformation at Interface and degradation in all-thin-film inorganic electrochromic device. Nanoscale. 2023

[27] Lopez R, Fuentes J, Gonzalez-Camps A, Benhaddouch T, Kaushik A, Metler CL, et al. Multimodal single-entity electrochemical fluoride sensor for fuel cell membrane degradation diagnostics. ECS Sensors Plus. 2022;**1**(3):035601

Section 4

Electrodeposition

Chapter 8

Electrodeposition of Metal Chalcogenide Thin Films

Arelis Ledesma Juárez, Alejandro Altamirano Gutiérrez and Arturo Fernández Madrigal

Abstract

Chalcogenide metals are compounds with optoelectronic characteristics that allow the construction of various optical and optoelectronic devices. The traditional preparation methods have been techniques such as reactive evaporation and sputtering. However, such methods have limited the construction of devices in large areas, particularly solar cells. Thin film solar cells based on Cu(In, Ga)Se_2 (CIGSe) absorbers have suitable optoelectronic properties. However, their commercial development is scarce since the efficiencies achieved on large surfaces differ from those obtained at the laboratory level. The electrodeposition technique is the most attractive from an economic point of view; however, high conversion efficiencies still need to be achieved with the latter. Several authors attribute this difference to different causes, such as using chemical additives in the preparation. The influence of the salt during its synthesis is not studied, so this work aims to use three different types of metal salts and voltages to produce (CIGSe) absorbing thin films by the electrodeposition technique. Also, the effect of nucleation type on two other substrates is studied. The various features of structural, morphological, atomic composition, and electrochemical characterization were to understand the formation, growth, and morphology of CIGS films, to obtain a suitable stoichiometry of thin film solar cells using this absorber.

Keywords: electrodeposition, solar cells, CIGS absorber, metal chalcogenide, X-ray diffraction

1. Introduction

Chalcogenide metals are materials composed of a chalcogen anion with a more electropositive element. They have been formed by those compounds based on sulfides, selenides, and tellurides. These materials contain various metallic minerals and can be used for many applications. Thin films of metal chalcogenide semiconductors find applications such as photovoltaics, batteries, electrocatalysis, or photonics. Examples of ternary chalcopyrite semiconductor materials such as CuZnSn$(S, Se)_2$ (CZT(S, Se) [1], CuInS_2, CuIn$(S, Se)_2$), or Cu(In, Ga)Se_2 (CIGSe) are among the most promising thin film materials for solar cell applications due to their properties and long-term stability, as well as adequate performance in low-light conditions [2], and suitable bandgap and absorption coefficient values [1, 3, 4]. Solar cells based on CIGSe films are the ones that have achieved high efficiencies, on the order of 23.35% using vacuum methods and 17.5% by the electrodeposition technique.

The electrodeposition technique is the most promising for preparing low-cost semiconductor solutions [5–7] and has been used to develop CdTe thin-film photovoltaic modules [8]. This technique employs the electro-reduction of ions to a specific reduction potential on a substrate to form thin films by applying an electric field. During this process, many parameters are involved, which influence the formation of a crystalline structure that allows achieving adequate growth of the thin film to form solar cells. Among the main parameters contributing to this type of film are the solution's pH, the deposition temperature, the additives added to the electrolytic baths, the position of the electrodes, the type of metallic salts used, the types of complexing agents, etc. These parameters mainly affect the type of nucleation, crystal size, electrical, optical, and optoelectronic properties, etc. For example, polyamide substrates treated with potassium fluoride as substrate obtained a cell with an efficiency of 20.4% [9]. Doping with rubidium fluoride (RBF) or cesium fluoride (CsF) improves the efficiency of solar cells by up to 20% [10]. Power has also increased by using Zn(O, S)-based window layers by 21% efficiency [11]. The addition of buffer solutions such as sodium sulfamate and potassium bi-phthalate helps stabilize the solution's pH, allowing Ga and ions to be incorporated into the films [12]. Using complexing agents has been an important strategy; thus, using KCN or potassium sodium potassium tartrate has allowed the control of Ga and ions into the films [13–15]. Some treatments applied to the coatings, through hydrogen and selenium, have increased efficiency by 9.4% [1]. It is essential to mention that incorporating In and Ga in the film is a determining factor in obtaining a film with an adequate structure, which allows the formation of solar cells with high conversion efficiencies. In this sense, this work describes the characteristics of CIGSe thin films using three types of metallic salts to prepare the electrolytic bath.

2. Experimental section

2.1 Materials and equipment

The co-electrodeposition technique uses three different salts to prepare the CIGSe thin films. **Table 1** shows the molar concentration of each metal salt used in the preparation of each bath. The purity and brand of each of the salts are copper (II) chloride dihydrate (99.99%), indium (III) chloride (99.999%), anhydrous gallium (III) chloride (99.99%), lithium chloride (≥99%), copper (II) nitrate hydrate (99.999%), indium (III) nitrate hydrate (99.999%), gallium (III) nitrate hydrate (99.9%), lithium nitrate (99.99%), copper (II) sulfate (≥99%), indium (III) sulfate hydrate (99.99%), gallium (III) sulfate (99.99%), selenious acid (99.999%). The

Bath No.	2.6 (mM)	4.5 (mM)	10 (mM)	8 (mM)	1 (mM)
1	$CuCl_2$	$InCl_3$	$GaCl_3$	H_2SeO_3	LiCl
2	$Cu(NO_3)_2$	$In(NO_3)_3$	$Ga(NO_3)_3$	H_2SeO_3	$LiNO_3$
3	$CuSO_4$	$In(SO_4)_3$	$Ga_2(SO_4)_3$	H_2SeO_3	Li_2SO_4

Table 1.
The molar composition of each of the salts used in the electrolytic baths.

chemicals were from Aldrich Chemical Co, the reagent lithium sulfate (99.7%), hydrochloric acid 37.4% ACS, nitric acid 70% ACS purchased from Fermont Company, and sulfuric acid 97.4% ACS from J.T. Baker, buffers pH three from Hydrion Company.

Two types of substrates use the first one made on Fluorine Tin Oxide glass, SnO_2:F (FTO) prepared by Delta Technologies. The Mo substrate on glass had designed in-house, using a Balzer BAE 250 direct current sputtering machine, using a Mo target with a diameter of 5.08 cm.

The thin films had prepared using a Bio-Logic SAS potentiostat model VSP s/n:0332 controlled with Ec-lab software. The films were analyzed using X-ray diffraction equipment using the diffractometer model DMAX-2200 with copper Kα radiation (λ_{Cu} = 1.5406 Å) at 40 kV, and the X-ray beam was at 0.5° grazing incidence (GIXRD). The atomic compositions of the films were obtained by EDS technique using Hitachi microscope model SU1510 SEM, which has a secondary electron detector and an energy-dispersive X-ray detector model INCA-x-act. An electron accelerating voltage of 8 kV and energy emission levels $K\alpha$ for copper, selenium, gallium, and $L\alpha$ for indium have been used to quantify the atomic composition. Atomic composition values were quantified over 600 μm x 600 μm. Micrographs of the films had obtained using a Hitachi model S-5500 SEM microscope with a secondary electron detector. An Alpha-step 100 profilometer has been used for thickness measurement. Diffuse reflectance spectra obtained by UV-VIS measure were performed using a Jasco V-670 spectrophotometer with a 2500-250 nm wavelength range 50 nm.

The absorption spectra had been obtained by the Kubelka-Munk equation [16].

$$\frac{K}{S} = \frac{(1 - R_\infty)^2}{2R_\infty} = F(R_\infty) \tag{1}$$

Eq. 1 shows the diffuse reflectance called the Kubelka-Munk function and the absorption coefficient and reflectivity, respectively. S and K are the so-called K-M scattering and absorption coefficients, respectively, and R_∞ = $R_{sample}/R_{standard}$. The band gap of thin films and the equivalent absorption coefficient had related through the following equation.

$$ahv = A\left(hv - E_g\right)^n \tag{2}$$

Eq. 2 [17] α is the linear absorption coefficient, A is an arbitrary constant, hν is the photon energy, and n equals 1/2 for direct transition-allowed materials.

When the incident radiation is perfectly diffusely scattered, the absorption coefficient K equals 2α. In this case, the Kubelka-Munk function is proportional to the absorption coefficient α; we obtain the relation.

$$[F(R_\infty)hv]^2 = (ahv)^2 \tag{3}$$

The Cyclic voltammetry (CV) and chronoamperometry (C.A.) had done in a three-electrode cell. The working electrode (WE) was FTO or Mo on glass (soda-lime glass) with a Mo film thickness between 1 and 2 microns and an active area of (1 cm × 3 cm), and the counter electrode (CE) was a platinum grid. Another platinum grid was a reference electrode (REF) for C. Studies, and Ag/AgCl electrode had used as the reference electrode for CV. The deposition potentials for the CIGS films were − 0.8, −0.9, and 1.0 V. The scan rates were 5 mVs^{-1}, 10 mVs^{-1}, and 20 mVs^{-1}.

3. Results and discussion

3.1 Atomic composition by EDS

Tables 2 and **3** show the atomic composition results obtained by the EDS technique. The number at the end of each sample indicates the applied potential; for example, number 1 corresponds to the application of −1.0 V. For 2 and 3, the applied voltages are −0.9 V and − 0.8 V, respectively. **Table 2** shows these results using Mo/glass substrates, and **Table 3** operates SnO_2:F (FTO) substrates at three different voltages. The thicknesses achieved for each sample are given in the same tables. The theoretical atomic percentages are Cu, 25%; In, 12.5%; Ga, 12.5%; and Se, 50% [18]; however, it has been found that to build more efficient solar cells, atomic ratios must reach values of 0.93 for [Cu]/[In+Ga] and 0.3 for [Ga]/[In+Ga] [19].

Table 2 shows that the samples obtained with electrolytic baths based on nitrate, sulfate, and chloride [Cu]/[In+Ga] values are high, higher than the desired theoretical value of 0.93. In contrast, the [Ga]/[In+Ga] values are similar to the theoretical value for all samples using the chloride bath and two of the nitrate samples (MN1 and MN2). These results indicate that the Ga content in these samples is adequate, but the Cu content is much higher than desired. This result has repeated for samples FC1, FC2, and FC3, but both values of the atomic ratios had reduced for the samples obtained with nitrate and sulfate baths using the FTO substrate, as shown in **Table 3**.

3.2 X-ray diffraction

Figure 1(a) and **(b)** show the diffractograms of the sample in **Table 1**, using electrolyte solutions of chlorides, nitrates, and sulfates, at potentials of −1.0, −0.9, or − 0.8 V, respectively, on two different substrates, Mo and FTO. **Figure 1(a)** shows a diffraction peak at 2Θ = 40, with the assigned plane (110) corresponding to Mo,

Samples	% At.						Thickness (μc)
	Cu (%)	In (%)	Ga (%)	Se (%)	Cu/(In + Ga)	Ga/(In + Ga)	
Mo- Chlorides							
MC1	26.90	15.55	6.10	51.44	1.24	0.28	1
MC2	28.11	15.83	5.48	50.60	1.31	0.25	1
MC3	29.13	12.78	5.39	52.70	1.60	0.29	0.9
Mo-Nitrates							
MN1	40.82	7.71	3.11	48.36	3.77	0.29	0.7
MN2	38.82	9.81	3.96	47.40	2.81	0.30	1
MN3	20.77	30.2	1.43	47.60	0.65	0.04	1
Mo-Sulfates							
MS1	27.18	22.86	1.26	48.7	1.03	0.04	1.1
MS2	42.20	12.67	0.74	44.13	3.14	0.05	2.4
MS3	41.56	14.03	1.38	43.03	2.76	0.07	1.7

Table 2.
Values of atomic percentages and composition ratios of Ga/(In+Ga) and Cu/(In+Ga), obtained by EDS of CIGS films, using Mo as substrate.

Sample	% At.						Thickness (μc)
	Cu (%)	In (%)	Ga (%)	Se (%)	Cu/(In + Ga)	Ga/(In + Ga)	
FTO-Cloruros							
FC1	26.16	14.46	6.14	53.22	1.27	0.30	0.7
FC2	38.40	7.23	4.78	49.59	3.24	0.39	1
FC3	31.07	11.97	5.01	51.86	1.82	0.30	0.9
FTO-Nitrates							
FN1	17.53	17.34	2.60	62.29	0.88	0.13	0.8
FN2	13.40	23.12	2.20	61.30	0.53	0.09	0.9
FN3	14.64	17.07	3.56	64.73	0.71	0.17	0.8
FTO-Sulfates							
FS1	17.11	32.36	0.49	50.78	0.53	0.01	0.8
FS2	24.69	29.93	3.29	42.10	0.74	0.09	0.7
FS3	23.69	25.81	1.12	49.39	0.88	0.04	0.7

Table 3.
Atomic composition percentages and Ga/(In+Ga) and Cu/(In+Ga) composition ratios obtained by EDS of CIGS films using FTO as substrate.

which has a cubic structure with lattice parameters, and whose lattice parameters are a = b = c = 3.1472 Å, according to PDF#42–1120. However, in **Figure 1(b)**, no diffraction peak related to the substrate type is observed because the X-ray beam scanning has been performed at a low angle, showing the diffraction planes of the film. **Figures 1(a)** and **1(b)** show that most of the samples have three principal planes identified: planes (112), (220), and (116), which coincide with the PDF #35–1102 structure and correspond to $CuGa_{0.3}In_{0.7}Se_2$. The above planes are practically the same for all samples prepared on Mo and FTO substrate, regardless of the type of electrolytic bath.

The crystal sizes for the samples shown in **Table 4** had been obtained from the Scherrer equation [20, 21]. **Table 4** shows that the values range from 4 to 15 nm. However, no increase or decrease in their values is observes with different voltages or the type of electrolytic bath.

3.3 Scanning electron microscopy

Figures 2(a,b) shows the morphology of CIGS samples obtained by scanning electron microscopy (SEM) technique, prepared on Mo and FTO substrates, using three types of metallic salts in the electrolytic bath (chlorides, nitrates, and sulfates), at three different voltages. SEM observed the growth of cauliflower-type formation, and clusters are formed by grains whose sizes vary from 0.05 to 0.2 microns. The dimensions of the grains change according to the type of metal salt used, the voltage applied, and the type of substrate. For example, in **Figure 2(a)**, the morphology of CIGS films on a Mo substrate has been observed. Samples MC1, MC2, and MC3, which had been prepared by applying a voltage of −1.0, −0.9, and − 0.8 V, respectively, show very similar granular growth in all of them, with values of 0.12, 0.125, and 0.16 microns, the size of these grains increasing as the applied voltage decreases. Similar behavior occurs when nitrate and sulfate salts are used, whereas granular growth

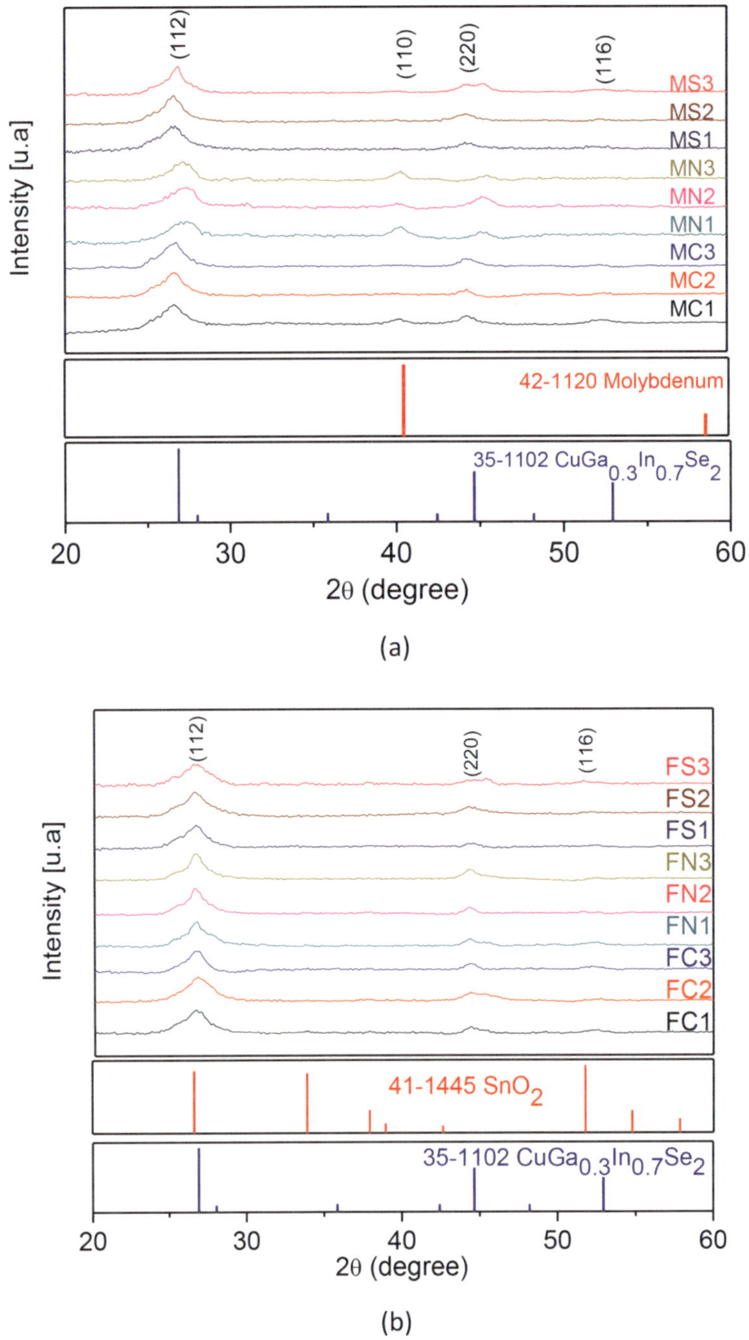

Figure 1.
(a) Shows the X-ray diffraction spectra of CIGS films deposited on Mo, and (b) shows the X-ray diffraction spectra of CIGS films deposited on FTO.

observes a relatively homogeneous distribution. In particular, MS3 shows the formation of flakes instead of spherical grains, like the rest of the samples. This flake formation indicates the formation of Cu-Se [22].

Mo substrate		FTO substrate	
Sample	**Crystal size (nm)**	**Sample**	**Crystal size (nm)**
MC1	8.08	**FC1**	9.02
MC2	10.84	**FC2**	4.61
MC3	7.48	**FC3**	10.55
MN1	15.68	**FN1**	9.27
MN2	6.23	**FN2**	10.67
MN3	10.35	**FN3**	9.48
MS1	7.61	**FS1**	9.76
MS2	6.91	**FS2**	8.37
MS3	6.93	**FS3**	4.81

Table 4.
Crystal size calculated from the Scherrer equation, using the highest intensity plane of each sample and principal planes of each of the sample.

Granular growth is observed in samples MS1 and MS2, which form cauliflower-shaped clusters. The development occurs by the clustering of grains on the surface. **Figure 2(b)** shows the morphology of the CIGS samples on an FTO substrate. These samples have a similar morphology to the samples on a Mo substrate. The average grain size ranges from 0.07 to 0.26 microns. Samples FC1, FC2, and FC3 have smaller grains than samples MC1, MC2, and MC3. Samples are obtained with a nitrate bath present small grains that agglomerate to form larger clusters, as shown by the morphology of samples FN1 and FN3, since FN2 presents the formation of flakes whose dimensions are up to 0.26 microns. Samples FS1, FS2, and FS3, formed by clusters of grains whose average size is between 0.16 and 0.21 microns, also show the formation of voids between these clusters.

3.4 J-t transient

The growth of CIGSe films is directly related to the applied overpotential; that is, the development of these films is by the kinetic mechanism of nuclei growth. Therefore, the nucleation and growth mechanisms during the co-deposition of CIGSe films can be studied using the models of the instantaneous and progressive nucleation equations Scharifker and Hill [23] have proposed. Eqs describe these models (4) and (5), respectively, where i_{max} and t_{max} are the peak current and t corresponds to the peak current in the CTT (Current-Time Transient) growth and nucleation region, respectively.

Progressive nucleation in 3D

$$\left(\frac{i}{i_{max}}\right)^2 = \left(\frac{1.2254}{\frac{t}{t_{max}}}\right) \times \left[1 - exp\left(-2.3367\left(\frac{t}{t_{max}}\right)^2\right)\right]^2 \tag{4}$$

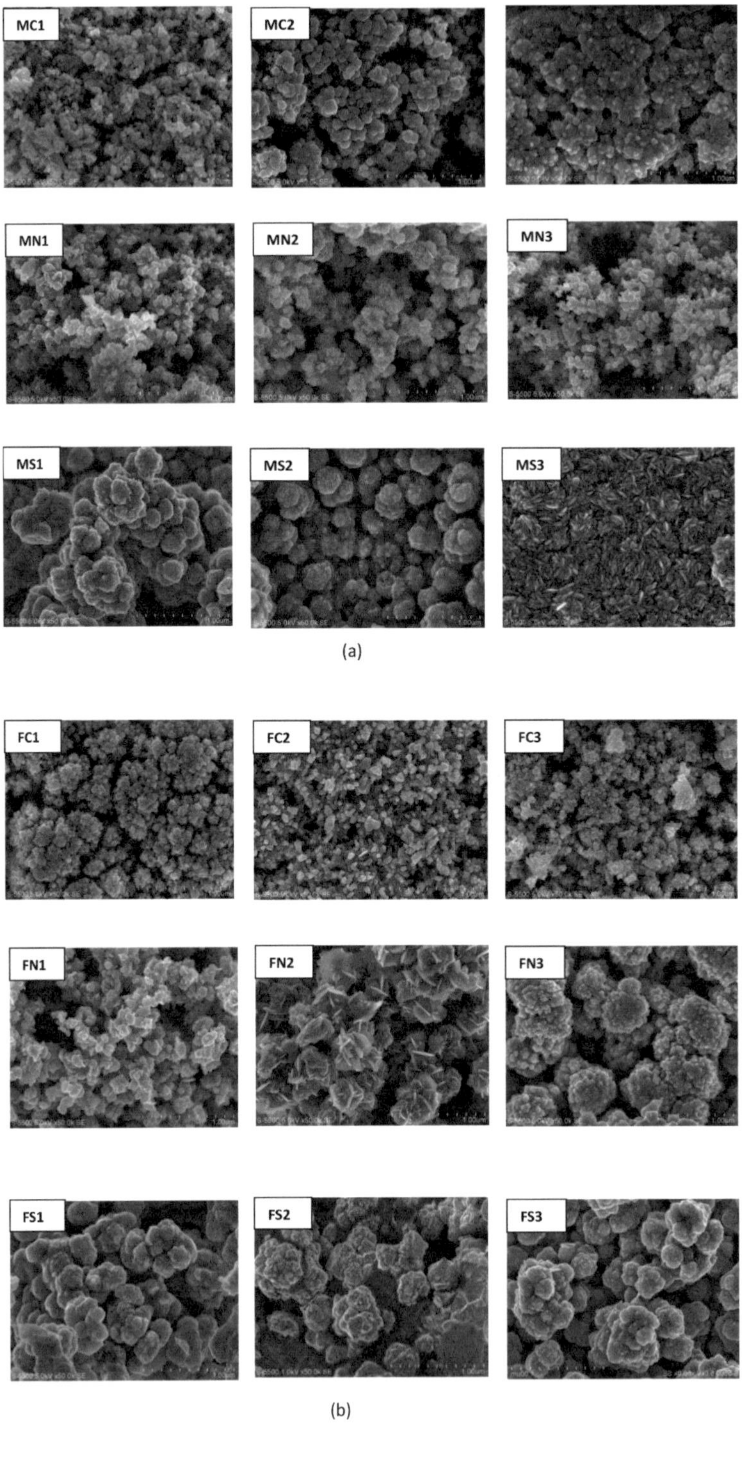

Figure 2.
Micrographs of CIGS films, fabricated using Mo (M) and FTO (F) as substrates, various types of salts (chlorides (C), nitrates (N), or sulfates (S)) using sulfate salts at overpotentials of -1.0 *V (1),* -0.9 *V (2), or* $-$ *0.8 V (3).*

3D instantaneous nucleation

$$\left(\frac{i}{i_{max}}\right)^2 = \left(\frac{1.9542}{\frac{t}{t_{max}}}\right) \times \left[1 - exp\left(-1.2564\left(\frac{t}{t_{max}}\right)\right)\right]^2 \tag{5}$$

Figure 3 shows the potentiostat current-time plots obtained during the electrodeposition of CIGSe films.

This behavior is associated with an increase in the electroactive area and the stabilization and growth of new nuclei; subsequently, the slow decay of the current observer indicates a mass transfer-controlled process. The transients show the characteristic behavior of diffusion-controlled three-dimensional (3D) nucleation processes [24]. **Figure 3** shows the transients of CIGSe film samples on Mo substrates (**Figure 3(a–c)**) at three different applied overpotentials. **Figure 3(d–f)** shows CIGS films on FTO substrates and the three overpotentials used. Each figure shows the plots obtained by Eqs. (4) and (5), where the rapid and progressive 3D theoretical growths can be seen. It observed that for all samples, the values obtained with these equations show a behavior very close to the 3D-instantaneous change, which is favored to a greater extent in FTO substrates than in molybdenum/glass substrates. F. Liu et al. report the same type of nucleation and growth of CIGS films as presented in this work, using only an electrolytic bath of chloride salts [25]. According to our results, using electrolyte baths other than chlorides produces fast 3D growth.

3.5 Optical properties

Table 5 shows the values obtained for the received bandwidth from the plots of $(F(R_\infty)^*h\nu)^2$ versus hν for the CIGS films prepared on molybdenum (Mo) and FTO. (Where $(F(R_\infty)^*h\nu)^2$ is Kubelka-Munk function and hν is the photon energy. As can be seen, the colored film on FTO has slightly higher bandwidth values than those deposited on Mo. These values are higher since the overpotential is higher for the films deposited on Mo and using chloride and sulfate electrolytic baths, but this is not the case for the nitrate ones. The results show that the value of the band gap in some films reported in the literature is about 1.14 eV, which suffers variations depending on the synthesis method; when comparing this value with those reported in **Table 5**, the values in this table are close.

3.6 Electrochemical evaluation

Figure 4(a) shows a potential reduction peak probably due to gallium deposition, at −0.726 V vs. Ag/AgCl potential, at a sweep speed of 5 mV/s, since the elements of the precursor salts of copper chloride, indium chloride, and lithium chloride are not deposited. On the other hand, at the same sweep speed, an anodic potential is seen at approximately −0.350 V vs. Ag/AgCl due to gallium's oxidation. Furthermore, a potential crossover shows at approximately −0.360 V due to the formation of a deposit on the Mo electrode surface, possibly gallium. At the rates of 10 mV/s and 20 mV/s, no anodic or cathodic potential peaks are observed, so, likely, the Mo electrode is not stable in the presence of the chlorides coming from the precursor salts at these two sweep rates in an interval of −0.350 to −1.0 V vs. Ag/AgCl a pseudocapacitive zone.

In **Figure 4c**, two consecutive potential peaks had been observed at approximately −0.002 and − 0.05 V vs. Ag/AgCl, respectively, which probably attributed to Cu reduction on the Mo substrate at sweep speeds of 10 and 20 mV/s. An anodic potential peak appears at approximately 0.35 V and 0.75 V vs. Ag/AgCl, probably due to the oxidation of the Cu film formed. However, the pH is acidic, it is very likely that Ga,

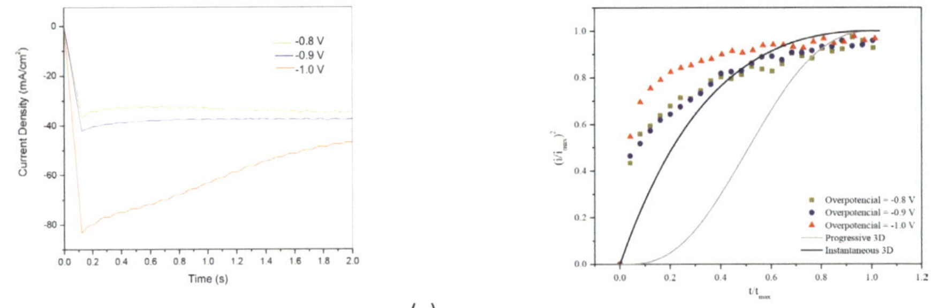

(a)

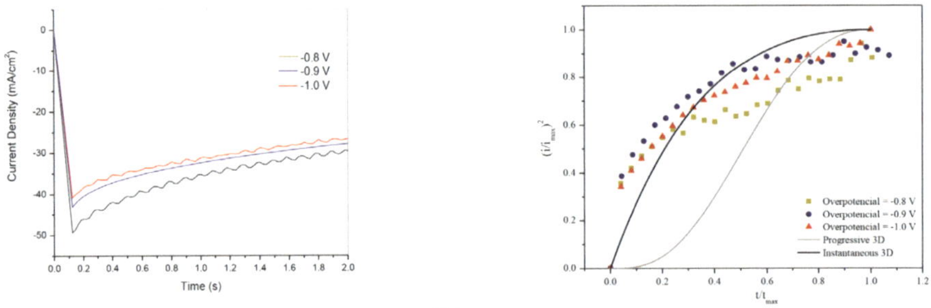

(b)

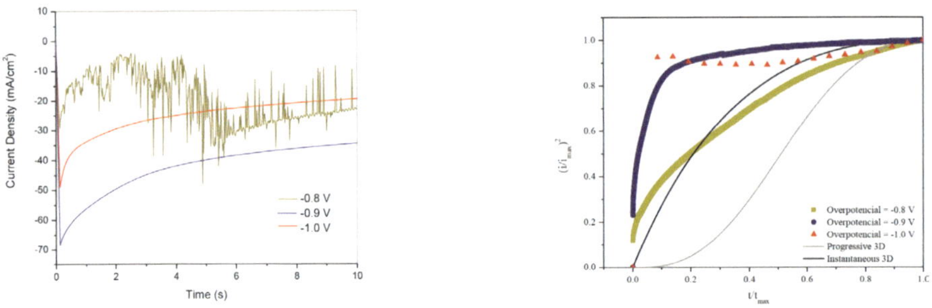

(c)

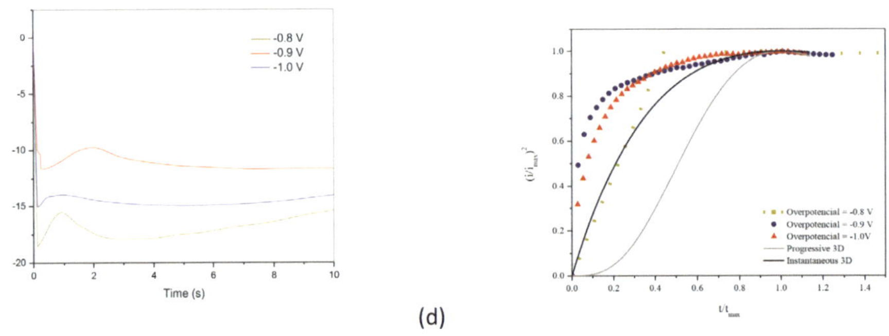

(d)

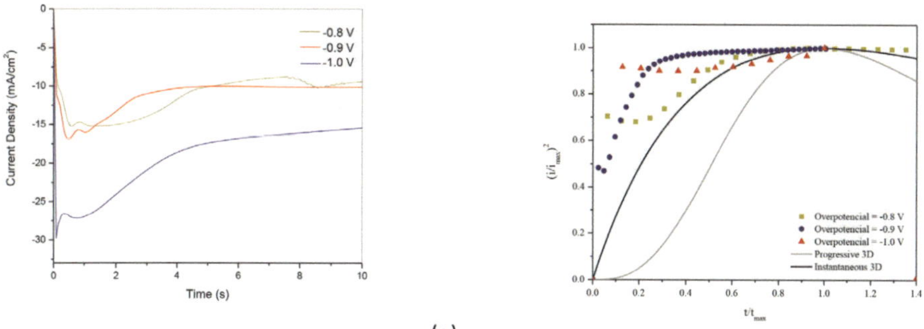

(e)

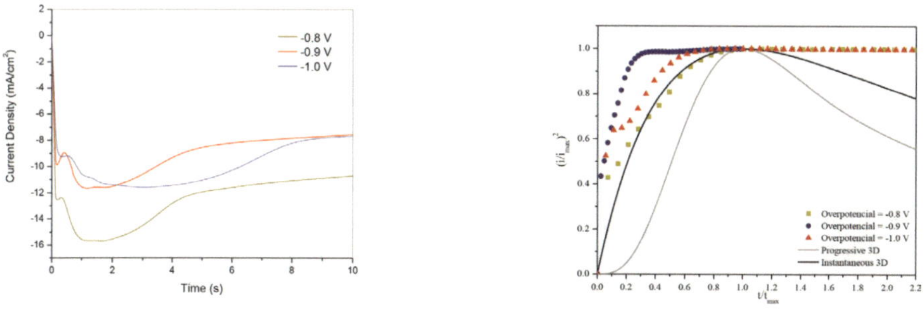

(f)

Figure 3.
(a), (b), (c) correspond to CIGS time-current transients using molybdenum/glass as substrate and chloride, nitrate, and sulfate salts, respectively. (d), (e), (f) correspond to the time-current transients of CIGS using FTO as substrate and chloride, nitrate, and sulfate salts, respectively.

Sample	Band gap value (eV)	Sample	Band gap value
MC1	1.27	FC1	1.02
MC2	1.32	FC2	1.42
MC3	1.47	FC3	1.34
MN1	1.24	FN1	1.33
MN2	1.30	FN2	1.27
MN3	1.32	FN3	1.34
MS1	1.24	FS1	1.28
MS2	1.17	FS2	1.26
MS3	1.09	FS3	1.32

Table 5.
*Bandwidth values for the CIGS films made with the electrolytic baths had shown in **Table 1**.*

In, and Li hydroxides of the precursor salts are starting on the electrode surface, based on Duchatelet A., Sidali T., et al. (2013). Consequently, this process does not allow these elements to deposit on the electrode. On the other hand, at the rate of 5 mV/s, no potential peaks are present.

Figure 4(e) shows three potential peaks associated with the reduction processes of the probable Cu, In, and Ga elements, at potentials of 0.012, −0.096, and − 0.205 V vs. Ag/AgCl, respectively, at the rate of 5 mV/s, and at this same rate an anodic peak is present due to an oxidation process at approximately −0.307 V vs. Ag/AgCl. However, at the velocities of 10 and 20 mV/s, neither cathode nor anodic peaks are

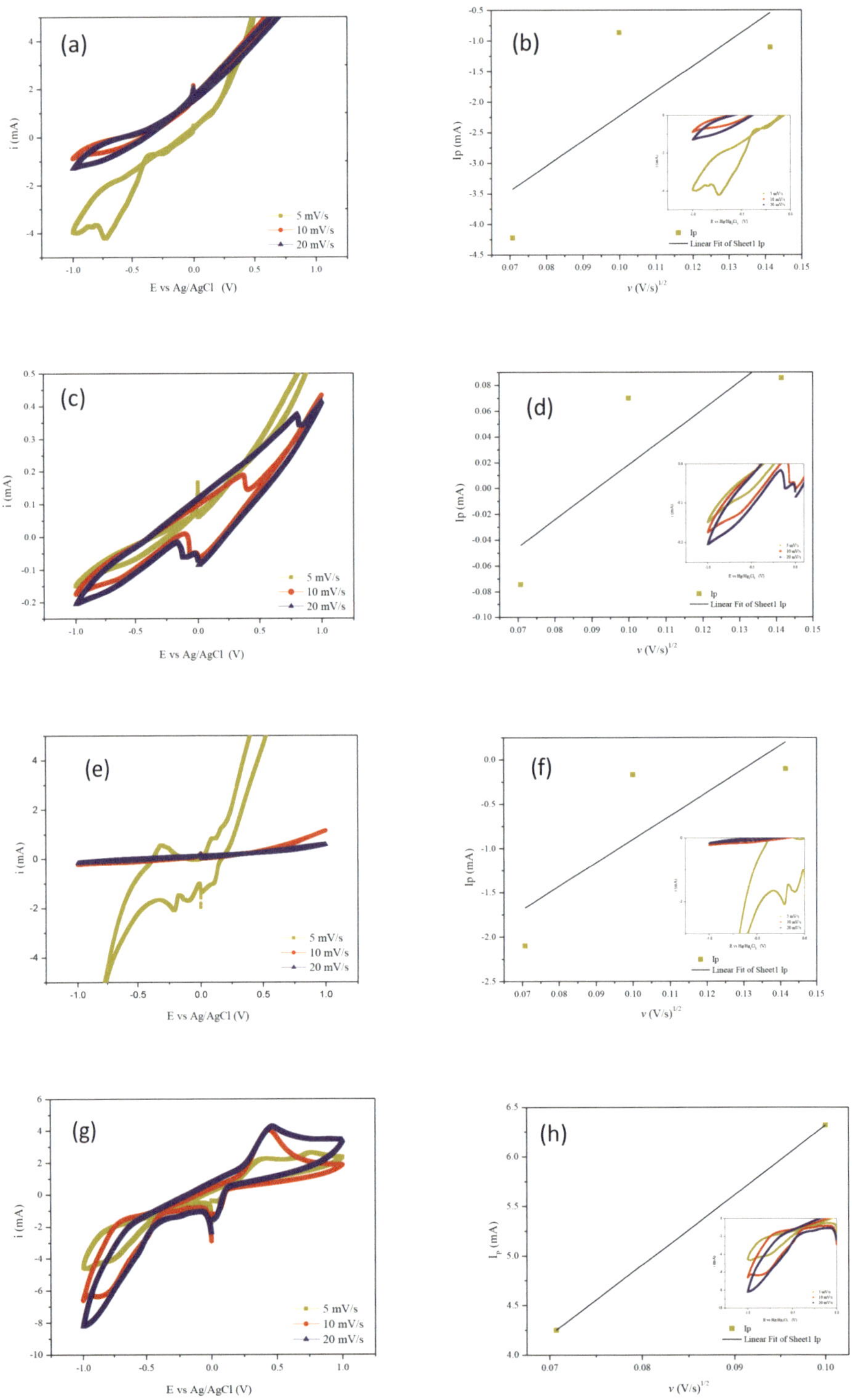

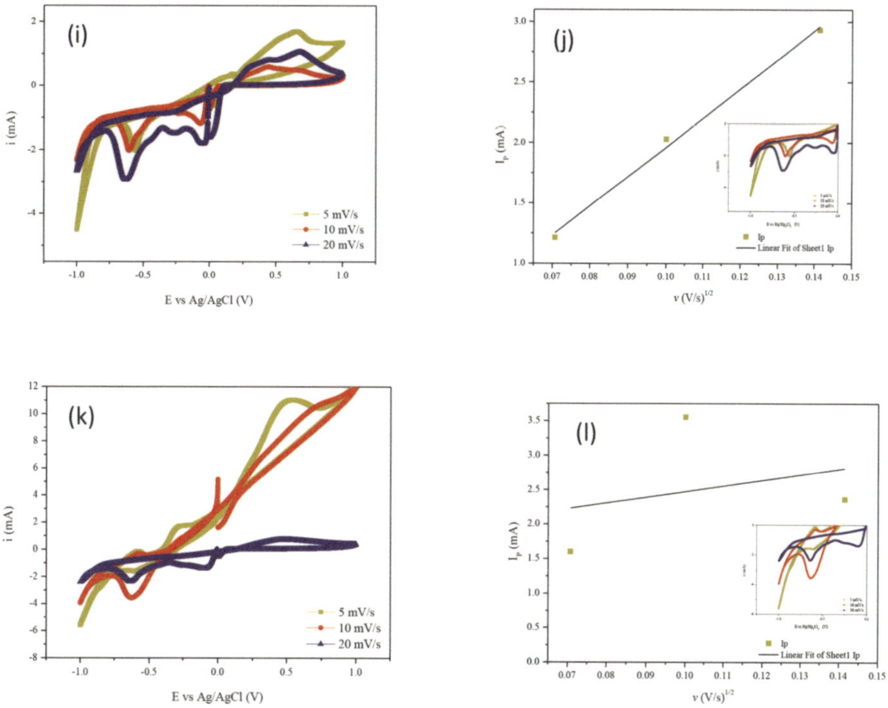

Figure 4.
Plots of the CIGS films deposited on molybdenum (a, c, e) or FTO (g, i, k), using chloride, nitrate, and sulfate salts, respectively, at overpotentials of −1.0 V (red), −0.9 V (wine), and −0.8 V (brown) each. (b, d, f, h, j, l) show the graphs between the maximum current and the square root of the scanning speed of each of the sales respectively.

observed; this is probably because the deposited Mo film was unstable in the presence of sulfates.

Figure 4(g) shows a potential peak corresponding to a reduction process around −0.003 V vs. Ag/AgCl, probably due to Cu reduction. Likewise, an oxidation process at about 0.456 V vs. Ag/AgCl is possibly due to Cu. These processes appear at all three sweep rates 5, 10, and 20 mV/s, which means that the FTO electrode surface is stable in the presence of chlorides of the precursor salts.

Figure 4(i) shows three potential peaks corresponding to the reduction around −0.055, −0.254, and − 0.600 V vs. Ag/AgCl, most probably deposited Cu, Ga, and In, respectively, and three oxidation processes observed at approximately 0.166, 0.432, and 677 V vs. Ag/AgCl, attributed to Cu, Ga, and In elements. These processes appear at the three scan rates of 5, 10, and 20 mV/s, which means that the FTO electrode surface is more stable than Mo in the presence of nitrates from the precursor salts.

Figure 4(k) shows two potential peaks mainly corresponding to reduction at about −0.083 and − 0.644 V vs. Ag/AgCl, most probably deposited Cu, Ga, respectively, and two oxidation processes observed at about −0.590 and 0.542 V vs. Ag/AgCl, attributed to Cu, Ga elements. These processes appear at all three sweep rates, 5, 10, and 20 mV/s, which means that the FTO electrode surface is less stable in the presence of sulfates than in nitrates.

Figure 4 show the plots between the peak current and the square root of the scan rate of each of the sales, respectively, which generally follow a straight-line trajectory confirming that the electrodeposition process is diffusion controlled and irreversible [26]. Furthermore, this process had shown in films deposited on the FTO substrate (h, j, l).

4. Conclusions

The results conclude that the CIGSe films fabricated using chloride baths and employing Mo and FTO as a substrate have values similar to the ideal Ga/(In+Ga) ratio of 0.3. However, when using nitrate and sulfate-based baths, this ratio has values close to perfect equilibrium only for samples MN1 and MN2. The rest of the pieces have low values. It indicates that chloride-based baths favor incorporating Ga into the film to form an ideal relationship with the composition. The analysis performed by X-ray diffraction found that the main phase corresponds to $CuGa_{0,3}In_{0,7}Se_2$ (PDF #35–1102). Whose principal diffraction planes are (112), (220), and (116), which have been identified in all the elaborated samples, independently of the type of electrolytic bath used or the type of substrate. It is important to note that all diffractograms do not show adequate polycrystallinity, which attribute to the kind of growth. The growth of the CIGS films is cauliflower type for all samples, observing an increase in size for an applied deposition voltage, that is, lower using chlorides to higher using sulfates. According to the models, the nucleation process of the fabricated films is instantaneous with 3D growth, which is more noticeable when FTO had used as substrate. When Mo substrate uses, a smaller band gap of CIGS films is getting; these values increase when the applied over potential also increases when chloride and sulfate salts have used. According to the graphs of maximum peak current versus the square root of the sweep speed for each metal salt used, a linear trend with a positive slope observes; that is, there is a directly proportional correlation between these two parameters, according to the model of the current equation and the sweep speed for cyclic voltammetry.

Acknowledgements

This work was supported through the PAPIIT-IN111320 project and the CONACyT Postdoc grant. We also thank Maria Luisa Ramón Garcia for XRD measurements, Rogelio Moran Elvira for SEM measurements, and José Campos Álvarez for EDS assistance.

Author details

Arelis Ledesma Juárez[1], Alejandro Altamirano Gutiérrez[2] and Arturo Fernández Madrigal[1*]

1 Instituto de Energías Renovables, Universidad Nacional Autónoma de México, Morelos, México

2 Centro Universitario de Tonalá, Universidad de Guadalajara, Jalisco, México

*Address all correspondence to: afm@ier.unam.mx

References

[1] Regmil G, Ashok A, Chawla P, Semalti P, Velumani S, Shailesh N, et al. Perspectives of chalcopyrite-based CIGSe thin-film solar cell: A review. Journal of Materials Science: Materials in Electronics. 2020;**31**:7286-7314. DOI: 10.1007/s10854-020-03338-2

[2] Green MA, Dunlop ED, Levi DH, Ebinger H, Yoshita J, Ho-Baillie AW, et al. Solar cell efficiency tables (version 54). Progress in Photovoltaics. 2019;**27**: 565-575. DOI: 10.1002/pip.3371

[3] Nakamura M, Yamaguchi K, Kimoto Y, Yasaki Y, Kato T, Sugimoto H. Cd-free Cu(In, Ga)(Se, S)2 thin-film solar cell with record efficiency of 23.35%. IEEE Journal of Photovoltaics. 2019;**9** (6):1863-1867. DOI: 10.1109/ JPHOTOV.2019.2937218

[4] Powalla M, Paetel S, Ahlswede E, Wuerz R, Wessendorf CD, Friedlmeier TM. Thin-film solar cells exceeding 22% solar cell efficiency: An overview on CdTe-, Cu(In, Ga)Se2-, and perovskite-based materials. Applied Physics Reviews. 2018;**5**:041602. DOI: 10.1063/ 1.5061809

[5] Pandey RK, Sahu SN, Chandra S. Handbook of Semiconductor Electrodeposition. New York, Basel, Hong Kong: Marcel Dekker, Inc.; 1996

[6] Lincot D, Guillemoles JF, Taunier S, Guimard D, Sicx-Kurdi J, Chaumont A, et al. Chalcopyrite thin film solar cells by electrodeposition. Solar Energy. 2004;**77** (6):725-737. DOI: 10.1016/j. solener.2004.05.024

[7] Lincot D. Electrodeposition of semiconductors. Thin Solid Films. 2005; **487**(1–2):40-48. DOI: 10.1016/j. tsf.2005.01.032

[8] Johnson DR. Microstructure of electrodeposited CdS/CdTe cells. Thin Solid Films. 2000;**361-362**:321-326. DOI: 10.1016/S0040-6090(99)00779-8

[9] Chirilă A, Reinhard P, Pianezzi F, Bloesch P, Uhl AR, Fella C, et al. Potassium-induced surface modification of Cu(In, Ga)Se2 thin films for high-efficiency solar cells. Nature Materials. 2013;**12**:1107. DOI: 10.1038/nmat3789

[10] Reinhard P, Bissig B, Pianezzi F, Avancini E, Hagendorfer H, Keller D, et al. Features of KF and NaF postdeposition treatments of Cu(In, Ga) Se2 absorbers for high-efficiency thin film solar cells. Chemistry of Materials. 2015;**27**:5755-5764. DOI: 10.1021/acs. chemmater.5b02335

[11] Friedlmeier TM, Jackson P, Bauer A, Hariskos D, Kiowski O, Wuerz R, et al. Improved photocurrent in Cu(In, Ga) Se2 solar cells: From 20.8% to 21.7% efficiency with CdS buffer and 21.0% Cd-free. IEEE Journal of Photovoltaics. 2015;**5**:1487-1491. DOI: 10.1109/ JPHOTOV.2015.2458039

[12] Liu J, Liu F, Lai Y, Zhang Z, Li J, Liu Y. Effects of sodium sulfamate on electrodeposition of Cu(In, Ga)Se2 thin film. Journal of Electroanalytical Chemistry. 2011;**1**(65):191-196. DOI: 10.1016/j.jelechem.2010.10.021

[13] Muthuraj JJJ, Rasmussen DH, Suni II. Electrodeposition of CuGaSe2 from thiocyanate-containing electrolytes. Journal of the Electrochemical Society. 2011;**158**:D54. DOI: 10.1149/1.3519997

[14] Stratieva N, Tzvetkova E, Ganchev M, Kochev K, Tomov I. Structural and optical properties of electrodeposited CuInSe2 layers. Solar Energy Materials

& Solar Cells. 1997;**45**:87-96. DOI: 10.1016/S0927-0248(96) 00070-0

[15] Aksu S, Wang J, Basol BM. Electrodeposition of In–Se and Ga–Se thin films to prepare CIGS solar cells. Electrochem. Solid State Lett. 2009;**12**: D33. DOI: 10.1149/1.3079481

[16] Morales AE, Mora ES, Pal U. Use of diffuse reflectance spectroscopy for optical characterization of un-supported nanostructures. Revista Mexicana de Fisica. 2007;**53**(5):18-22

[17] Abdullahi SS, Güner S, Koseoglu Y, Musa IM, Adamuy BI, Abdulhamid MI. Sımple method for the Determınatıon of the band gap of a Nanopowdered sample Usıng Kubelka Munk theory. Journal of the Nigerian Association of Mathematical Physics. 2016;**35**:241-246

[18] Farooq L, Alraeesi A, Zahmi SA. A review on the electrodeposition of CIGS thin-film solar cells. IEOM Society International. 2019;**26**(28):158-185. Corpus ID: 223560420

[19] Eberspacher C, Fredric C, Pauls KS. Thin-film CIS, alloy PV materials, fabricated using non-vacuum, particles-based techniques. Thin Solid Films. 2001;**387**:18-22. DOI: 10.1016/S0040-6090(00)01729-6

[20] Lee H, Yoon H, Ji H, Lee D, Lee J, Yun J, et al. Fabrication of CIGS films by electrodeposition method for photovoltaic cells. Journal of Electronic Materials. 2012;**41**(12):3375-3381. DOI: 10.1007/ s11664-012-2252-x

[21] Holzwarth U, Gibson N. The Scherrer equation versus the Debye-Scherrer equation. Nature Nanotechnology. 2011;**6**:534. DOI: 10.1038/nnano.2011.145

[22] Saïdia H, Ben Alayaa C, Boujmila MF, Durand B, Lazzari JL, Bouaïcha M. Physical properties of electrodeposited CIGS films on crystalline silicon: Application for photovoltaic hetero-junction. Current Applied Physics. 2020; **20**:29-36. DOI: 10.1016/j.cap.2019.09.015

[23] Pourbaix M. Atlas of Electrochemical Equilibria in - Aqueous Solutions. Houston, Texas: National Association of Corrosion Engineers; 1974. p. 77084

[24] Scharifker B, Hills G. Theoretical and experimental studies of multiple nucleation. Electrochimica Acta. 1983; **28**:879-889

[25] Gupta A, Srivastava C. Nucleation and growth mechanism of tin electrodeposition on graphene oxide: A kinetic, thermodynamic and microscopic study. Journal of Electroanalytical Chemistry. 2020;**861**:113964 (1-8. DOI: 10.1016/j.jelechem.2020.113964

[26] Liu F, Huang C, Lai Y, Zhang Z, Li J, Liu Y. Preparation of Cu (In, Ga) Se_2 thin films by pulse electrodeposition. Journal of Alloys and Compounds. 2011; **509**:L129-L133. DOI: 10.1016/j. jallcom.2010.12.031